MY ARCTIC JOURNAL

A YEAR AMONG ICE~FIELDS AND ESKIMOS

1891–1892

MY ARCTIC JOURNAL

A YEAR AMONG ICE-FIELDS AND ESKIMOS

1891–1892

by

Josephine Diebitsch-Peary

WITH AN ACCOUNT OF
THE GREAT WHITE JOURNEY
ACROSS GREENLAND

BY

ROBERT E. PEARY
CIVIL ENGINEER, U.S. NAVY

NEW INTRODUCTION
BY **ROBERT M. BRYCE**

Cooper Square Press

Published by Cooper Square Press
An Imprint of the Rowman & Littlefield Publishing Group
150 Fifth Avenue, Suite 817
New York, New York 10011
www.coopersquarepress.com

Distributed by National Book Network

Library of Congress Cataloging-in-Publication Data Available

⊖™ The paper used in this publication meets the minimum requirements of American National Standard for Information Sciences—Permanence of Paper for Printed Library Materials, ANSI/NISO Z39.48–1992.
Manufactured in the United States of America.

INTRODUCTION TO THE
COOPER SQUARE PRESS EDITION

By the standards of her day, Josephine Diebitsch-Peary led a very unusual life. As with most unusual lives, hers was the product of an unusual ambition. Although, like many Victorian women, it was her husband's ambition and not her own, her desire to be an active participant in its realization was what brought her to wider notice.

Josephine Diebitsch was born to German immigrant parents on May 22, 1863 in Washington, D.C. Her father was head of the interlibrary exchange at the Smithsonian Institution and Josephine became his assistant. At nineteen she met Robert E. Peary, a civil engineer in the U.S. Navy. Even their meeting was a product of Peary's budding ambition.

Peary had once been engaged to a hometown girl, but broke off the engagement when he moved to Washington to take up his appointment as a draftsman for the Coast and Geodetic Survey. After obtaining his Navy commission through a competitive exam, he aimed to rise high in Washington society and to make powerful friends. A product of provincial Maine, he made a conscious study of improving his social graces, including a course in ball-

room dancing. It was at a dance that he met Miss Diebitsch in 1882. This encounter began a six-year courtship,
interrupted by journeys in quest of Peary's fondest desire:
fame and an honorable name that would make him
known throughout the world.

At first, he hoped to find both in connection with his
work on the survey of a route for the Trans-Isthmian Ship
Canal across Nicaragua. Then Peary became interested in
Greenland's unexplored ice cap, which he penetrated 100
miles in 1886 on a self-financed excursion. When he had
to return to Nicaragua in 1887, he debated what he
should do about Jo: "That she loves me I know; that she
can make me happy I think; that she would hamper me
less than any woman I have met or am likely to meet I am
confident. [Still] I shrink from voluntarily chaining myself; & have to submit my last & fairest dream to the cold
light of prosaic daily life." Nevertheless, before Peary left,
he proposed, and they were married after his return, on
August 11, 1888. They honeymooned in Seabright, New
Jersey. Peary's overprotective mother accompanied them.
Jo naturally resented this intrusion and ever after felt herself second in Bert's affections.

Peary was assigned to the New York Naval Shipyard.
The couple set up housekeeping, first on Madison Avenue, then in a boardinghouse on 32nd Street. Jo was help-
lessly in love with the man whom she thought could not
be more handsome, and believed that, if not ahead of his

mother, she at least stood ahead of Bert's ambition. "Bert puts the Canal aside," she wrote to her mother, "and devotes the whole time to me, and if he is as happy as I am, then we are the happiest people in the world."

But he was not. "Prosaic daily life" soon lost its novelty; the old yearning for fame only lay dormant for awhile.

With the abandonment of the Nicaraguan canal route for one across Panama, the southern pathway to fame closed to him. Peary's thoughts turned north again. He began to promote an expedition, the object of which would be the crossing of Greenland's ice cap. When Norwegian explorer Fridtjof Nansen accomplished this feat, Peary proposed a survey of Northern Greenland to determine its extent in the direction of that much-coveted exploratory goal, the North Pole (Peary's own secret ambition). But for the time being, this ultimate destination remained only a dream, one that caused him many periods of self-doubt and despair.

In 1890, Peary's fortune changed when he was transferred to the League Island Naval Yard in Philadelphia. A lecture on his proposed expedition gained him the attention and support of the Academy of Natural Sciences of Philadelphia. Suddenly, it seemed as if his dreams might be within his grasp: "Fame, money & revenge goad me forward till sometimes I can hardly sleep with anxiety lest something happen to interfere with my plans," he confided to his mother.

Peary worked tirelessly, putting his expedition together on a shoestring and obtaining a leave of absence from his naval duties to execute his plans. He engaged young, eager amateurs as assistants and took care to hire no potential rival for attention or for credit upon his return. But as the expedition prepared to sail from Brooklyn on June 6, 1891, the center of attention was not Robert Peary, but his wife, who would be going as well.

The newspapers vacillated between condemning Robert Peary for exposing his young wife to the dangers of the frigid north, which had claimed many a hardy man, and commending Josephine Peary for her loyalty in agreeing to accompany her husband. The large crowd on the dock to see the *Kite* off was anxious to get a glimpse of this brave woman bound for the "icy Sahara."

My Arctic Journal, first published in 1893, is just that: a record of Jo Peary's experiences on her husband's North Greenland Expedition of 1891–1892. During research for my book, *Cook & Peary: The Polar Controversy, Resolved*, I always sought primary sources, including accounts by participants in the events that I wished to portray. Jo Peary's book was readily at hand, but I wondered how much of it reflected her actual experiences and feelings. Although complete diaries of some other expedition members were easily located in the vast Peary Family Collection housed at the National Archives, only parts of Mrs. Peary's diary could be found there. Promising leads in New England turned up none of the missing portions. However, a comparison of what survives shows that many important passages in her book—such as her description of the accident that broke her husband's leg on the way to Greenland—are reproduced from her original diary nearly verbatim. Of course, a large amount of material has been edited out: chiefly commonplaces, annoyances, and statements of a personal nature. But because Jo was unable to abandon completely her Victorian notions of cleanliness, civilization, and propriety—much less come to any real understanding of the "strange little brown people" she encountered in the Inuit, who flocked to see the wonders of the white men's house—*My Arctic Journal* remains intimate and strikingly frank, and can be said to embody the authentic voice of its author, despite the editing. In many ways, it is the best published record of the events of this expedition and gives an immediate sense of the novelty of her experience and of the drudgery that underlies all worthwhile accomplishments and seemingly easy successes.

It also gives some insight into the other expedition members, but Jo is far less frank in this regard. Each of the men who signed on with Peary was interesting in some way. Langdon Gibson was the brooding brother of the soon-to-be-famous creator of the Gibson Girl, Charles Dana Gibson. Of all the expedition members, Langdon alone had no interest in ever returning to the Arctic again. Eivind Astrup, who accompanied Peary on the latter's successful crossing of Greenland's ice cap in 1892, had the opposite reaction to his introduction to the Far North. He returned with Peary the next year, only to fall out with him and to form plans for his own expedition. But Astrup's life was cut tragically short when he froze on a skiing tour in his native Norway in 1896. John Verhoeff, an eccentric who also had future Arctic plans, lost his life in an accident just prior to the expedition's return to the U.S. Matt Henson, Peary's black manservant, who lately has risen highest in recognition as an explorer, receives little serious notice in Jo's pages. He was regarded merely as the expedition's cook and handyman at this time. The expedition's physician, Dr. Frederick A. Cook, became an explorer in his own right and, eventually, a figure surrounded by great controversy in the wake of his bitter dispute with Peary over who was first to the North Pole.

Jo left out of her book her candid opinions of each of these men:

> I hate the people who surround me more & more every day. Every action speaks of lowness & coarseness. I am afraid I am too much of an aristocrat to ever get used to such people as these. It is very evident to me that B-[ert] does not share this feeling & rather resents it in me. I can see plainly in all that I know of him that he has been brought up to always do & say things for the sake of what others may think or say while I have been accustomed to say what I thought & do as I wished without

caring a rap what any one said or thought. The best thing for me to do is to avoid all conversation with the boys; in that way I avoid offending B- by my speech. . . .

The members of the party as they appear to me. Dr. F. A. Cook, an exceedingly coarse man with not an idea of gentlemanly behavior. A former milk man who from his savings studied medicine, although he cannot write a page without misspelling words. How good a physician he is of course I do not know, he was kindness itself to us during Bert's unfortunate accident & a woman could not have been more gentle. This I appreciate highly & shall never forget but his uncouth manners & unrefined talk grate on me constantly. Imagine a physician . . . speaking of "belching up wind from the 'stommick' as a sign of fermentation in the intestines" at the table. When he has finished eating he rolls back on his chair & rubs his belly, frequently comes to the table in his dirty undershirt & pants, boasts that he only combs his hair on Sunday which not having been cut since he left N.Y. hangs in stringy masses over ears & neck. The other day while he was marking out something on the table cloth with his fingers a louse walked over his hand. Altogether he is a dirty specimen of manhood, but a good worker & a very good-hearted person.

J. M. Verhoeff comes next I suppose; he is an uncanny & very homely dwarf. Nothing gentlemanly about him. Has evidently had no "home training." He has some money & has had a course in a scientific school. He is very eccentrick & thinks himself a model of goodness & smartness. There is no doubt that he is not quite right in the "upper story." He thinks nothing of spitting on the floor & talks of his relatives, one is a policeman, another

keeps the poorhouse somewhere &c, &c. He confines himself to
2 meals a day & then just stuffs himself, putting everything on
his plate at the same time & mincing it all up. Anything but ap-
petizing for the rest of us. But the boys say he is very good-
hearted. He can never attend to more than one thing at a time
but this he does thoroughly & can be depended upon.

Next comes Langdon Gibson. There is nothing repulsive
about his looks as there is in the 2 former men. He was formerly
messenger boy & then clerk in Wall street & it is evident that he
has been brought in contact with a better class of people. His
appearance I am sorry to say is the only favorable thing about
him. He is lazy, shiftless, a flatterer & a thoroughly deceitful fellow, besides being a coward & though he claims to be an orni-
thologist he knows little more about the birds than the average
person who has gone gunning for sport. General education he
has none . . . & is continually exposing his ignorance.

Eivind Astrup being a foreigner & not able to always express
himself in English seems reticent & quiet when I happen to be
in the room so I can't say much of him. He would be a nice
bright looking boy if he kept himself clean. The others have
taught him to swear & he seems to think it is a big thing to swear
as much as possible. The first thing he teaches the natives when
they come here is "go to h—" He is still young & I *think* a gentle-
man born, so I have hopes of his improvement. These together
with the colored boy are the civilized men with whom I am
brought in daily contact. Bert thinks I ought to treat them as my
equals & see only their good points & feels hurt that I do not do
this; he has no idea what it costs me to even treat them with
ordinary politeness. I would much rather ignore them.

Some of the men harbored similar feelings about Jo
Peary. She was willing to do her share, but her mere presence was a source of tension, and was frequently the topic
of conversation among the unmarried men. (Only Henson, at this time, was married.) Although the argumenta-
tive Verhoeff agreed that a woman had no place in the
Arctic, he would not participate in some of the other top-
ics of discussion. "I did not like the smutty talk," he confided to his diary.

Verhoeff, who disdainfully called her "the Woman,"
and especially Gibson, took a dislike to both Pearys,
sensed her classist prejudices and his unwillingness to
share a particle of glory, and considered both of them im-
perious.

If there was an exception to this attitude, it was Dr.
Cook. Verhoeff believed that, despite what Jo wrote about
him in her diary, Cook was her favorite. Perhaps their
common German heritage accounted for this supposed
affinity, but it seems inescapable that Cook had a great
respect for Jo's courage and loyalty. Certainly, she was in
Cook's mind in 1901, when he wrote this odd conclusion
to his article, "A Walk to the North and South Poles":
"Perhaps some hardy American . . . will ere long arise
whose strong hand, or that of his wife perhaps, shall hoist
the . . . Stars and Stripes beyond the present borderland."

Except for the loss of Verhoeff, Peary's expedition was
a total success, which only fueled his further ambition. Im-

mediately, he laid plans for another, larger expedition and went on a strenuous lecture tour to finance it. He was so much away from home that Jo saw less of him than she had in the Arctic. When he gained leave to return to Greenland, the depression, anxiety, and homesickness that she had experienced that first Arctic winter did not deter her from resolving to go again, even though she was pregnant. But after a heart-to-heart talk with a confidential friend, Jo was convinced this would be foolhardy. She told Bert that she could not take the risk, but Peary seemed so shaken by this decision that she relented.

My Arctic Journal ends with a paragraph written on the shores of Inglefield Gulf, where she was about to spend her second Arctic winter. She says: "Everything points to the success which Mr. Peary hopes for. What the future will bring, however, no one can tell."

The second expedition started out well, but because of chance misfortune and Peary's own overreaching plans, his Second Greenland Expedition was as complete a failure as the first had been a success. The only bright spot came with the birth of Marie Ahnighito Peary on September 12, 1893. She was so fair that the Inuit wanted to touch her to see if she was warm, since to them it seemed she must be made of snow. Marie was known as "The Snow Baby" ever after.

Peary's poor planning left him critically short of supplies when severe weather in the spring of 1894 rebuffed his first attempt to cross the ice cap. As a result, he was forced to reduce his party to a bare minimum if he were to have any chance of success the next spring. Returning to America was not a problem for most of Peary's disgruntled assistants, who harbored on an even larger scale the same animosity that the first expedition's members had shown toward the Pearys on an even larger scale. Although one returning member described their experience as "a perfect Hell," the implication (in a letter sent back by Peary) that they had returned out of cowardliness hurt their pride and made Peary future enemies. But when Peary decided that even Jo had to return, only her brother, Emil, who had come on the relief ship, was able to convince her that she must for Marie's sake. Still, Jo was deeply hurt by what she considered her husband's lack of faith in her.

Once home, Jo determined to raise money for what she called "The Greenland Scientific Expedition of 1895," even though in reality its only purpose was to relieve Bert and to bring him back home for good. To this end she went on the lecture circuit under the management of the impresario of the lyceum, Major J. B. Pond, who felt she was "in possession of a speaking talent that would have made her a permanent success." From her proceeds and from solicitations of the scientific and geographical societies in New York, Philadelphia, and Washington, she was able to raise enough money to send a ship in the summer of

1895. With it, she sent Bert a heart-rending letter recall-
ing her bitter disappointment as they sat together by a
favorite little pond in a valley near their expedition hut.

> When you told me your plan a year ago last April as we sat near
> the shore of Baby Lake, I felt as if you had put a knife into my
> heart & left it there for the purpose of giving it a turn from time
> to time.
>
> I have reviewed our married life very carefully, my hus-
> band, & think I am resigned to the place which you gave me an
> hour after we were married. It was No. 2 & how it did hurt, has
> continued to hurt & will hurt until the end. A year ago you made
> it No. 3 . . .
>
> My Bert, my life, if you have not been successful won't you be
> content to put fame in the background & live for me a little, as
> you once did?

In the Arctic, Peary, who had barely escaped with his
life in his most recent attempt to repeat his crossing of
Greenland, also regretted his decision. At about the same
time that Jo was penning her ship-borne letter, he was
writing to Jo, even though he had no means of sending it:
"Never have I felt more lonely than tonight, never if God
grants me to take you in my arms again will I leave you
for so long again. The past runs black, the future blacker.
. . . I feel at times I am going mad. I have lost my sanguine

hope, my élan, I am an old man, I think at times I have lost you."

But even though Peary declared upon his return, "I shall never see the North Pole unless someone brings it here. I am done with it," he could not shake off his desire for the fame that he would achieve as its discoverer. Jo did her best to dissuade him, but he could not help himself, despite his vows. He had to return to the Arctic. "You are right dear, life is slipping away," he told her, "but there is something beyond me, something outside of me, which impels me irresistibly to the mark." And so Jo accompanied him again, but only for a summer attempt in 1897 to recover a giant meteorite at Cape York. Peary's success in securing it also secured his and Jo's financial future, when in 1909 she sold it to the American Museum of Natural History for the equivalent of $640,000 in today's money.

Jo would not return to the Arctic when Peary embarked on his projected five-year "siege" of the Pole in 1898, with the assistance of his newly formed cadre of wealthy backers incorporated as The Peary Arctic Club. She was pregnant again. But when she received word from the 1899 relief ship that her husband had lost most of his toes to frostbite, she decided that she must fetch Bert back once again and for all. She was free to go; her infant daughter had lived only seven months. She sailed with Marie on the *Windward* in the summer of 1900. When she arrived at

Payer Harbor on the shores of Ellesmere Island, she found that Bert was 250 miles farther north than the ship could penetrate. But she soon made another discovery that was far more disturbing.

In 1885, even before he had ever seen the Arctic, Peary had written in his diary: "It is asking too much of masculine human nature to expect it to remain in an Arctic climate enduring constant hardship, without one relieving feature. Feminine companionship not only causes greater contentment, but as a matter of both mental and physical health and the retention of the top notch of manhood it is a necessity. . . .

Let white men take with them native wives, then from that union may spring a race combining the hardiness of the mothers with the intelligence and energy of the fathers. Such a race would surely reach the Pole if their fathers did not succeed in doing it.

Jo was staggered not only to find that he had acted on his own advice and taken an Inuit mistress, "a creature scarcely human," but that he had also fathered a child by her, when Allakahsingwah, with innocent pride, showed her the son of *Pearyaksoah*. That the boy had lived while her own girl had wasted and died was almost too much for her to bear. "These are the darkest days of my life," she confided to a sympathetic fellow passenger. She then poured out her grief, horror, and anguish in a twenty-six-page letter that she planned to leave for her absent husband when the ship sailed: "To think she has been in your arms, has received your caresses, has heard your lo-vecries—I could die at the thought. . . . You gave me three years of the most exquisite pleasure that can be had; after that the pleasure was pretty evenly divided with the pain until now it is all pain, except the memory of what has been."

But the *Windward* did not sail. She was caught in the ice for the winter. When Peary was returning to Payer Harbor the following spring, he encountered two Inuit on the way north. He was surprised to learn that a ship was in Payer Harbor and shocked to be told that "Mitty Peary" was on board, as he knew nothing of the death of his second child. Then they handed him a copy of Jo's letter. After he had read it, he hesitated for several days, timing his arrival at the ship for his birthday, May 6, so as to soften the blow that he knew was coming.

Jo sailed that summer aboard the *Windward*, leaving her husband to his ambition. When she returned the next year, he had not only failed again, but also seemed crushed in spirit as well as impaired permanently by the loss of his toes. But once home, he started to plan yet again for one final chance to "make good."

In 1901, Jo had authored a children's book, *The Snow Baby*, about the birth of her child in the "Snowland" in-habited by the "strange little brown people." It also included the adventures of "Miss Bill," a twelve-year-old Inuit who accompanied her to the United States and lived with the Pearys in New York for a year. It outsold any of her husband's weighty narratives about his expeditions to the Arctic, and she followed this achievement in 1903 with a second book for young readers, *Children of the Arctic*, which also did well.

It seemed that Jo had finally resigned herself to her secondary position in Peary's personal universe, or perhaps she came to realize that as long as his ambition remained unfulfilled, her dreams would have to wait. But her daughter took up where she had left off by writing him a letter:

My dear, dear Father,

Of course I know the papers are not always right, but I read that the Peary Arctic Club are trying to get your consent to go north again. I think it a dog's shame. . . . I know you will do what pleases mother and me, and that is to stay with us at home.

I have been looking at your pictures . . . and I am sick of looking at them. I want to see my father. I don't want people to think me an orphan.

Please think it over.

Your loving,

Marie

Neither his daughter's pleading nor even the birth in 1903 of his much-desired son, Robert Emil Peary, could keep him at home, however. His millionaire backers provided him with a specially built ice-ship, the *Roosevelt*, with which Peary was sure he would succeed. He left once more for the Arctic in 1905. While he was away, Jo worked hard to secure for Bert the position that he had always coveted: the Chief of the Bureau of Yards and Docks, the only position in the Civil Engineering Corps that could hold the rank of admiral. She hoped that if he were appointed, he would never return to the Arctic again. But the unexpected vacancy was filled before Bert reappeared. When he did, he had failed again, and had battered his new ship nearly to pieces in the process. In frustration, Jo asked him, "Why don't you just tell them that you have found [the Pole] and let it go at that?"

"Because there are a hundred ways in which I would be found out," she said he replied. One thing that Jo may never have found out was that her husband had fathered another child in Greenland, though Peary hinted at it when he wrote her that Allakahsingwah had a new son, "a tall, fine looking fellow."

Peary's 1906 claim of the "Farthest North" achieved thus far earned him one last chance. That opportunity came in July 1908 with the blessings of Theodore Roosevelt: "I believe in you, Peary," said the president while touring his refurbished ship. "And I believe in your success—if it is in the possibility of man."

When Peary next returned on September 5, 1909,

claiming that success on April 6, 1909, he found the world acclaiming Frederick A. Cook as the discoverer of the North Pole instead of himself. Although Cook had only just arrived back from the Arctic on September 1 and was being honored in Copenhagen, Peary's rival claimed to have reached the Pole on April 21, 1908.

Despite this disappointing news, Jo Peary was ecstatic when she received a telegram from Labrador:

> **I HAVE THE D.O.P.** [A family euphemism meaning "Damned Old Pole"] **AM WELL. WILL WIRE AGAIN FROM CHA-TEAU. BERT.**

"I knew it! I knew it!" she cried. "I have been thinking all day that we should hear good news to-day. Oh, I am so glad, so glad!"

"You don't know what it means to me," she told a reporter. "Twenty-three years he has been working for this, and during all this time I have just existed, that's all. Hardly a year of real happiness during all that time have I had because of the worry and anxiety and fear. But it is all over now. No one knows how much my husband has suffered or sacrificed to reach the pole. Many times he has risked his life. The best years of his life have been spent in that frozen country, far removed from every comfort and from everything. It has been his life's work. He has studied and labored for success, and now that it has come no one can appreciate how happy it makes me."

Then came the inevitable question about Bert's rival. "I
don't want to talk about Dr. Cook's discovery. There has
been too much said about that, and I don't want to com-
ment on it. Whatever I might say would be misunder-
stood, anyway, and people are so unkind about such
things. If he reached the pole ahead of Mr. Peary it makes
no difference.

"We shall begin life again when Mr. Peary gets home. I
shall not hear of his going north again."

Privately, however, she did not believe that Cook had
won, and referred to him in her letters to Bert only as
"that creature." In October 1907, when it had first become known that Cook had stayed north and would try
for the Pole, Bert and she considered it a joke. When they
learned more of Cook's serious preparations for an attempt on the Pole from his backer, John R. Bradley, Peary
cried foul. He publicly condemned Cook for usurping
"his" Eskimos, dogs, and plans. But it wasn't long before
Peary had even darker thoughts, deciding that his rival's
intention might be to make a false claim to the goal that
had evaded his own best efforts for all those years. Soon
after Peary reached Labrador, he accused Cook of "hand-
ing the world a gold brick."

The tension between Cook's announcement and Bert's
telegram was more than Jo could stand. Upon hearing
of her husband's presence in Labrador, she packed the
children off to Sydney, Nova Scotia, to greet him person-

ally. She waited impatiently for weeks for him to come down from the North, and when she heard that the *Roosevelt* was finally approaching, she could wait no longer. When she transferred from the yacht that carried her and the children out to the *Roosevelt*, Peary received a kiss from Marie and a smart salute from Robert. He immediately retired to his cabin with Jo, and in twenty minutes, emerged in so exuberant a mood that he caught his daughter about the waist and waltzed her around the deck.

Initially, things went badly for Peary in his struggle with Cook. His attacks on his rival made him look like a very bad loser, while Cook's gentlemanly demeanor in the face of them won the doctor much sympathy. Eventually, however, with the help of Peary's well-healed and influential backers, Cook's claim was gradually brought into public doubt. Then the committee of the University of Copenhagen, to which Cook had submitted his "proofs" that he had attained the Pole, rejected his claim, saying they contained no evidence that he had actually reached it. In the meantime, a committee of the National Geographic Society—for self-serving reasons and without much serious scientific consideration—had certified Peary's claim. To the public, not knowing the lack of rigor exercised by the society's committee, Peary seemed to have passed the scientific examination that Cook had failed so ignominiously, and was proclaimed the true discoverer of the North Pole.

But although Peary now had his fame, his efforts to discredit Cook had erased most of his own popularity with the public, who considered him an unworthy hero, even if he had won the prize. Cook's discredit also raised serious doubts in many people's minds about Peary, whose narrative of his polar conquest had very much in common with Cook's rejected one. This made some wonder if Peary had any more proof of his achievement than Cook.

Nevertheless, the National Geographic Society's certification cleared the way for Peary to capitalize on his victory over Cook. In 1910, Peary toured Europe with Jo and Marie—picking up an armload of medals, trophies, and honorary degrees—but his attempt to be retired with the rank of rear admiral led to congressional scrutiny of his polar records and raised serious doubts about his own claim, forever casting a shadow over the honorable name that he so desired.

"No one will ever know how the attack on my husband's veracity affected him, who had never had his word doubted in *any* thing at *any* time in his life," Jo Peary said. "He could not believe it. And the personal grilling which he was obliged to undergo at the hands of Congress . . . hurt him more than all the hardships he endured in his sixteen years of research in the Arctic regions and did more toward the breaking down of his iron constitution than anything experienced in his explorations."

For a few years Jo Peary finally got the only thing that she had ever wanted. The couple divided their time between summers at their cottage on Eagle Island in Casco Bay, and winters at their comfortable home in the Cleveland Park neighborhood of Washington, D.C. Book and magazine contracts, lectures, and endorsements netted Peary what would equal millions of today's dollars, and there was another triumphant foreign tour. But the coming of World War I all but ended public interest in the North Pole and who had discovered it. With the outbreak of hostilities in Europe, Peary devoted himself to preparing America for the coming war effort and took a particular interest in coastal air defenses.

After two years of acute illness, Robert Peary died of pernicious anemia on February 20, 1920. He was buried as a national hero in Arlington and the National Geographic Society erected a monument over his grave identifying it as the resting place of the "Discoverer of North Pole."

Yet, the questions already raised about the authenticity of his claim persisted and grew more numerous. The opening of Peary's personal papers in 1984 did not put the doubts and suspicions to rest, as his advocates hoped, but instead proved that Peary had little beyond his bare word to back his assertion that he actually achieved his life's ambition. All evidence indicated that Peary had taken Jo's advice: he just said he had discovered the North Pole when his last opportunity to do so was a failure like all the rest. One by one, the hundred ways in which he would be found out came to light. But Jo always believed. Her life had been so wedded to her husband's ambition, what else could she do? To think otherwise would have made all of her sacrifices in vain. After her husband's death, Jo retired from public life and was rarely seen, but stood ever ready to defend him from behind the scenes whenever his reputation or deeds were questioned.

One of her last public acts, in 1955, was to give the taffeta flag that she had made for him in 1898 to the National Geographic Society in recognition of its steadfast friendship. Peary had kept it wrapped around his body on his expeditions—except to clip bits from it to mark significant milestones in his career, including a long diagonal swatch left at the spot that he claimed to be the North Pole. At age ninety-two, and with less than a year to live, Jo Peary was too frail to make the presentation herself, so her daughter did the honors. "I always have felt that my father would never have discovered the North Pole if it hadn't been for mother," Marie told the society's officers. "She could have said, 'We are only young once; why should we be separated?' "

Jo *had* said that—many times. "Life is almost over," she had often told Bert, "and we have missed most of it." But he never listened.

Life was nearly new, however, when she wrote *My Arctic Journal*, and all Jo's regrets were still in that unknown future, which she had written about on the shores of Inglefield Gulf in 1893. She was then only thirty and by the side of a man whom she loved with great passion, and for whom she would have sacrificed much, and, as it turned out, did sacrifice nearly everything to Peary's peculiar ambition.

In the hundred years since her book was last in print, much has changed about the world's view of her husband, but Jo Peary's story endures. It is one of steadfast loyalty and faithfulness despite everything that was still to come—or that slipped away, unexperienced.

ROBERT M. BRYCE
Monrovia, Maryland
March 2001

Robert M. Bryce is the author of *Cook & Peary: The Polar Controversy Resolved*. He is widely regarded as the leading authority on the controversy surrounding the rival claims of Frederick Cook and Robert Peary to have been the first man to reach the North Pole. He has been a scholar of the subject for more than twenty-five years and has studied extensively the personal papers of each of the explorers. During research for his book, Mr. Bryce read all of the extent letters of Josephine D. Peary at the National Archives in Washington, D.C. He also read the surviving portions of her original 1891–1892 diary, on which *My Arctic Journal* is based, which are part of the Peary Family Collection housed there. Mr. Bryce has previously written the introductions to Cook's *My Attainment of the Pole*, Peary's *The North Pole*, and Matthew Henson's *A Negro Explorer at the North Pole*, all available from Cooper Square Press.

INTRODUCTORY NOTE

On June 6, 1891, the steam-whaler "Kite," which was to bear the expedition of the Philadelphia Academy of Natural Sciences northward, set sail from the port of New-York, her destination being Whale Sound, on the northwest coast of Greenland, where it had been determined to pass the winter, preliminary to the long traverse of the inland ice which was to solve the question of the extension of Greenland in the direction of the Pole. The members of the expedition numbered but five besides the commander, Mr. Peary, and his wife. They were Dr. F. A. Cook, Messrs. Langdon Gibson, Eivind Astrup, and John T. Verhoeff, and Mr. Peary's faithful colored attendant in his surveying labors in Nicaragua, Matthew Henson. This was the smallest number that had ever been banded together for extended explorations in the high Arctic zone. A year and a quarter after their departure, with the aid of a relief expedition conducted by Professor Angelo Heilprin, Mr. Peary's party, lacking one of its members, the unfortunate Mr. Verhoeff, returned to the American shore. The explorer had traversed northern Greenland from coast to coast, and had added a remarkable chapter to the history of Arctic exploration.

The main results of Mr. Peary's journey were:

The determination of the rapid convergence of the shores of Greenland above the 78th parallel of latitude, and consequently the practical demonstration of the insularity of this great land-mass;

1

The discovery of the existence of ice-free land-masses to the northward of Greenland; and

The delineation of the northward extension of the great Greenland ice-cap.

In the following pages Mrs. Peary recounts her experiences of a full twelvemonth spent on the shores of McCormick Bay, midway between the Arctic Circle and the North Pole. The Eskimos with whom she came in contact belong to a little tribe of about three hundred and fifty individuals, completely isolated from the rest of the world. They are separated by hundreds of miles from their nearest neighbors, with whom they have no intercourse whatever. These people had never seen a white woman, and some of them had never beheld a civilized being. The opportunities which Mrs. Peary had of observing their manners and mode of life have enabled her to make a valuable contribution to ethnological learning.

THE PUBLISHERS.

PREFACE

This plain and simple narrative of a year spent by a re-
fined woman in the realm of the dreaded Frost King has
been written only after persistent and urgent pressure from
friends, by one who shrank from publicity, and who reluctantly yielded to the idea that her experiences might be of
interest to others besides her immediate friends.

I have been requested to write a few words of introduction; and while there may be some to whom it might occur
that I was too much interested to perform this task properly,
it must nevertheless be admitted that there is probably no
one better fitted than myself to do it. Little, indeed, need
be said.

The feeling that led Mrs. Peary through these experiences
was first and foremost a desire to be by my side, coupled with
the conviction that she was fitted physically as well as otherwise to share with me a portion at least of the fatigues and
hardships of the work. I fully concurred in this feeling, and
yet, in spite of my oft-expressed view that the dangers of life
and work in the Arctic regions have been greatly exaggerated,
I cannot but admire her courage. She has been where no
white woman has ever been, and where many a man has hesitated to go; and she has seen phases of the life of the most

northerly tribe of human beings on the globe, and in many ways has been enabled to get a closer insight into their ways and customs than had been obtained before.

I rarely, if ever, take up the thread of our Arctic experiences without reverting to two pictures : one is the first night that we spent on the Greenland shore after the departure of the "Kite," when, in a little tent on the rocks—a tent which the furious wind threatened every moment to carry away bodily—she watched by my side as I lay a helpless cripple with a broken leg, our small party the only human beings on that shore, and the little "Kite," from which we had landed, drifted far out among the ice by the storm, and invisible through the rain.   Long afterward she told me that every unwonted sound of the wind set her heart beating with the thoughts of some hungry bear roaming along the shore and attracted by the unusual sight of the tent; yet she never gave a sign at the time of her fears, lest it should disturb me.

The other picture is that of a scene perhaps a month or two later, when—myself still a cripple, but not entirely helpless—this same woman sat for an hour beside me in the stern of a boat, calmly reloading our empty firearms while a herd of infuriated walrus about us thrust their savage heads with gleaming tusks and bloodshot eyes out of the water close to the muzzles of our rifles, so that she could have touched them with her hand, in their efforts to get their tusks over the gunwale and capsize the boat.   I may perhaps be pardoned for saying that I never think of these two experiences without a thrill of pride and admiration for her pluck.

In reading the pages of this narrative it should be remembered that within sixty miles of where Kane and his little party endured such untold sufferings, within eighty miles of where Greely's men one by one starved to death, and within less than fifty miles of where Hayes and his party and one portion of the "Polaris" party underwent their Arctic trials and tribulations, this tenderly nurtured woman lived for a year in safety and comfort: in the summer-time climbed over the lichen-covered rocks, picking flowers and singing familiar home songs, shot deer, ptarmigan, and ducks in the valleys and lakes, and even tried her hand at seal, walrus, and narwhal in the bays; and through the long, dark winter night, with her nimble fingers and ready woman's insight, was of inestimable assistance in devising and perfecting the details of the costumes which enabled Astrup and myself to make our journey across the great ice-cap in actual comfort.

Perhaps no greater or more convincing proof than this could be desired of what great improvements have been made in Arctic methods. That neither Mrs. Peary nor myself regret her Arctic experiences, or consider them ill-advised, may be inferred from the fact that she is once more by my side in my effort to throw more light on the great Arctic mystery.

R. E. PEARY,
Civil Engineer, U. S. N.

FALCON HARBOR, BOWDOIN BAY,
GREENLAND, August 20, 1893.

TABLE OF CONTENTS

CHAPTER I

NORTHWARD BOUND

First Sight of Greenland — Frederikshaab Glacier — Across the Arctic Circle —
Perpetual Daylight — Sunlit Disko — The Climb to the Ice-cap — Dinner at
Inspector Anderssen's — A Native Dance — From Disko to Upernavik — Upernavik — The Governor and his Wife — The Duck Islands — Gathering Eggs
and Eider-down and Shooting Ducks.

Wednesday, June 24. We have sailed and tossed, have
broken through the ice-barriers of Belle Isle Straits, and once
more ride the rolling swells of the broad Atlantic. Our three
days' jam in the ice has given us a foretaste of Arctic naviga-
tion, but the good little " Kite" speeds northward with a confidence which inspires a feeling of security that not even the
famed "greyhounds of the ocean" afford. Genial Captain
Pike is on the bridge and off the bridge, and his keen eye is
casting for the land. When I came on deck to-day I found
the bold, wild coast of Greenland on the right. It was a
grand sight — the steep, black cliffs, some of them descending
almost vertically to the sea, their tops covered with dazzling
snow, and the inland ice flowing through the depressions
between their summits; at the foot of the cliffs gleamed bergs
of various sizes and shapes, some of them a beautiful blue,
others white as snow. The feature of the day was the

Frederikshaab glacier, which comes down to the sea in latitude 62° 30′. It did not, however, impress me as being very grand, owing perhaps to our being so far from it. Its face is seventeen miles long, and we could see it like a wall of white marble before us. Long after we had passed it, it still appeared to be with us, and it kept us company nearly all day. Just beyond the glacier was disclosed the most

Out on the Billowy Sea.
The First Fragment of Greenland Ice.

beautiful mountain scenery imaginable. The weather was deliciously warm, and revealed to us a new aspect of Arctic climate. It seems strange to be sitting on deck in a light coat, not even buttoned, and only a cap on my head, in the most brilliant sunshine, and gazing on snow-covered mountains.

Thursday, June 25. We were promised another lovely day, but after noon the weather changed and a cool wind sprang up, which helped to push our little craft along at a good rate. To-night we shall have the midnight sun for the first time, and it will be weeks, even months, before he sets for us again. Everything on deck is dripping from the fog which has gathered about us.

Friday, June 26. In spite of the thick fog we have been making good time, and expect to be in Disko, or more properly Godhavn, about noon to-morrow. We saw our first eider-ducks to-day. Numerous bergs again gleam up in the distance, probably the output of the Jakobshavn glacier.

Capt. Richard Pike — "On Duty."

Tuesday, June 30. We have been in a constant state of excitement since Saturday morning, when we first set foot on Greenland's ice-bound shores. The pilot, a half-breed Eskimo, came on board and took us into the harbor of Godhavn shortly after nine o'clock. Mr. Peary, Captain Pike, Professor Heilprin, and myself went ashore and paid our respects to Inspector Anderssen and his family. They were very attentive

to us, and invited " Mr. and Mistress Peary " to stay with them
during their stop in Godhavn — a pleasure they were, however,
compelled to forego.   In the afternoon a party of us from the
"Kite" set out on our first Arctic tramp, our objective point
being the summit of the lofty basalt cliffs that tower above
the harbor.   My outfit consisted of a red blanket combination
suit reaching to the knee, long knit stockings, a short eider-
down flannel skirt reaching to the ankles, and the "kamiks," or
long-legged moccasins, which I had purchased in Sidney.
The day was exceptionally fine and sunny, and we started off
in the best of spirits.   Never had I seen so many different
wild flowers in bloom at once.   I could not put my foot down
without crushing two or three different varieties.   Mr. Gibson,
while chasing a butterfly, slipped and strained the cords of
his left foot so that he was obliged to return to the ship.
Never had I stepped on moss so soft and beautiful, all shades
of green and red, some beds of it covered so thickly with tiny
pink flowers that you could not put the head of a pin down
between them.   We gathered and pressed as many flowers as
we could conveniently carry — anemones, yellow poppies,
mountain pinks, various *Ericaceæ*, etc.   Sometimes our path
was across snow-drifts, and sometimes we were ankle-deep in
flowers and moss.   Mountain streams came tumbling down in
every little gully, and their water was so delicious that it
seemed impossible to cross one of these streams without
stooping to drink.   Our advance was very slow, as we could
not resist the temptation of constantly stopping to look back

and feast upon the beauties of the view. Disko Bay, blue as sapphire, thickly studded with icebergs of all sizes and beautifully colored by the sun's rays, lay at our feet, with the little settlement of Godhavn on one side and the brown cliffs towering over it. As far as the eye could reach, the sea was dotted with icebergs, which looked like a fleet of sail-boats. The scene was simply indescribable. We reached the summit, at an elevation of 2400 feet, and built a cairn, in which we placed a tin box containing a piece of paper with our names written upon it, and some American coins. From the summit of these cliffs we stepped upon the ice-cap, which seemed to roll right down to their tops. The temperature was 91° F. in the sun, and 56° in the shade. As we descended a blue mist seemed to hang over that part of the cliffs that lay in shadow, and the contrast with the white bergs gleaming in the sapphire waters below was very striking. We returned to the foot of the cliff after eight o'clock. On Sunday we made another expedition, to the Blaese Dael, or "windy valley," where a beautiful double waterfall comes tumbling through the hard rock, into which it has graven a deep channel. We gathered more flowers, and collected some seaweed; the mosquitos, of which we had had a foretaste the day before, were extremely troublesome, and recalled to memory the shores of New Jersey. When we reached the first Eskimo hut, a number of the piccaninnies[1] came to me and presented me

---

1 The Eskimos frequently designate their children as piccaninnies, a word doubtless introduced by the whalers.

with bunches of wild flowers. We gave them some hardtack
in return, and they were happy.

Mr. Peary, Professor Heilprin, myself, and two other members of our party dined with the inspector in the evening,
joining some members of the Danish community, who had
also been invited. The course consisted of fresh codfish with
caper-sauce, roast ptarmigan, potatoes boiled and then
browned; and for dessert, "Rudgrud," a "dump," almonds,
and raisins. There was, following European custom, a varied accompaniment of wines.

After dinner the gentlemen went up-stairs to examine the
geological and oölogical collections of the inspector, while the
ladies preferred the parlor with their coffee. Were it not for
the outer surroundings, it would have been difficult to realize
that we were in the distant Arctic realm, so truly homelike
were the scenes of the little household, and so cheerful the
little that was necessary to make living here not only comfortable, but pleasant. The entire community numbers barely
120 souls, nine tenths of whom are Eskimos, mainly half-
breeds; the remainder are the Danish officials and their
families, whose recreation lies almost entirely within the little
circle which they themselves constitute.

Toward nine o'clock we visited the storehouse, where a
native ball was in progress. Several of our boys went the
rounds with the Eskimo "belles," but for me the odor of the
interior was too strong to permit me to say that looking on
was an "unalloyed pleasure." The steps were made to the

music of stringed instruments, over which the resident half-breeds have acquired a fair mastery. The participants and onlookers were all in a lively frame of mind, but not uproarious; and at the appointed time of closing—ten o'clock—all traces of hilarity had virtually been banished.

We had hoped to leave early on the following morning, but it was not until near two o'clock that the fog began to lift,

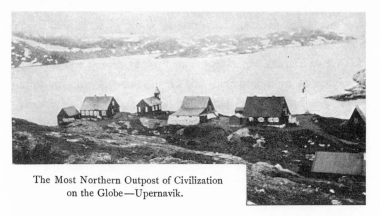

The Most Northern Outpost of Civilization
on the Globe—Upernavik.

and that a departure was made possible. Firing the official salute, and dipping our colors, we gave three hearty cheers in honor of our first Greenland hosts, and sailed out of the rock-bound harbor. It soon cleared up, and we were able to make our normal seven knots an hour. This morning it was foggy for a while, but it cleared up beautifully, and now we are just skimming along, and expect to reach Upernavik, the most northern of the Danish settlements in Greenland, about nine o'clock in the evening.

Thursday, July 2. We did not reach Upernavik until 2.30
yesterday morning, owing to a very strong current which was
running against us all the way from Godhavn. We remained
up all night, and at 1.30 A. M. were enjoying the dazzling
brightness of the sunshine. Mr. Peary took a number of
photographs between midnight and morning. Upernavik is a
very different-looking place from Godhavn. There are four
frame-houses and a little church. The natives live in turf huts,
very miserable-looking habitations, built right down in the
mud. As soon as our ship steamed into the harbor, in
which two Danish vessels were at anchor, the governor, Herr
Beyer, came on board with his lieutenant-governor, a young
fellow who had arrived only three days before. We returned
the visit at noon, and were pleasantly received by the governor and his wife, a charming woman of about thirty years,
who had been married but a year, and whose fondness for
home decoration had expressed itself in the pictures, bric-à-
brac, fancy embroideries, and flowering plants which were
everywhere scattered about, and helped to make up an extremely cozy home. As in all other houses in the country,
the guests were treated to wine immediately on entering, and
with a delicate politeness the governor presented me with a
corsage bouquet of the flowers of Upernavik, neatly tied up
with the colors of Denmark. Our visit was fruitful in the
receipt of presents, among which were Eskimo carvings, a
dozen bottles of native Greenland beer, and a box of "goodies,"
addressed to "Miss Peary," and to be opened, as a reminder,

on Christmas eve. The hospitality shown to us could not have been more marked had our friendship extended over many years.

Our visit was a brief one, as we were to weigh anchor early in the afternoon. We steamed away from Upernavik and headed north. The fog had cleared away and disclosed a giant mountain towering above us in the harbor. The sun shone brightly, and the sea was smooth as glass and blue as turquoise. The night promised to be a beautiful one, but I resisted the temptation to stay up, having been up the entire night before, and the greater part of the one before that. At 4 A. M. Captain Pike knocked at our door and informed us that in half an hour we would be at the Duck Islands. Here we were to land and all hands shoot eider-ducks and gather their eggs for our winter supply. We were soon on shore, and then began a day's sport such as I had often read about, but never expected to see. The ducks flew in thick flocks all about us, and on every side were nests as large as a large hen-nest, made of eider-down and containing from three to six eggs. The nests were not hidden, but right out on the rocks in full sight. Alas! we were too, late ; the ducks were breeding, and out of 960 eggs we did not get over 150 good ones. As I had not taken my gun, I spent the time in gathering down, and collected forty-three pounds in five hours. After returning to the "Kite" for breakfast, we visited a second island, and there I bagged a bird, much to my satisfaction. Altogether ninety-six ducks were shot.

2

CHAPTER II

IN THE MELVILLE BAY PACK

Melville Bay — On the Edge of the Dreaded Ice-pack — Fourth of July — Butting the Ice — Accident to the Leader of the Expedition — Gloom on the " Kite "— Blasting the " Kite " out of a Nip — A Real Bear and a Bear Hunt — A Chase on the Ice — A Phantom Ship — Free of the Pack and in the North Water at Last — The Greenland Shore to Barden Bay — First Sight of the Arctic Highlanders.

Thursday, July 2. We are opposite the "Devil's Thumb," latitude 74° 20′, and now, at 8 P. M., are slowly making our way through the ice which marks the entrance into the Melville Bay "pack."

Friday, July 3. At midnight the engine was stopped, the ice being too thick for the "Kite" to make any headway. At 6.30 A. M. we started again, and rammed our way along for an hour, but were again forced to stop. At eleven o'clock we tried it once more, but after a couple of hours came to a standstill. We remained in this condition until after five o'clock, when the engine was again started. For two hours we made fairly good progress, and we thought that we should soon be in open water, but a small neck of very heavy ice stopped us. While we were on deck, the mate in the "crow's-nest," which was hoisted to-day, sang out, "A bear! A bear!" Off in the distance we could see an object floating, or rather

swimming, in the water, and in a minute the boys were climbing helter-skelter over the sides of the " Kite," all with guns,
although some soon discovered that theirs were not loaded;
but the bear turned out to be a seal, and
not one of about thirty shots hit him.
It is now nearly 11 P. M. The sun is
shining beautifully, and it is perfectly
calm. I have worn only a gray spring
jacket, which I have found sufficient for
the balmy temperature. At midnight
the cannon was fired, the flags were
run up and dipped, and the boys fired
their rifles and gave three cheers for
the Fourth of July. The thermometer
marked 31°.

Saturday, July 4. The ice remains
stubborn, and we are fast bound. All
around the eye sees nothing but the
immovable pack, here smooth as a table,
at other places tossed up into long hum-
mock-ridges which define the individual
ice-cakes. Occasional lanes of water

" A Bear! A Bear! "

appear and disappear, and their presence gives us the one
hope of an early disentanglement. The event of the day has
been a dinner to Captain Pike, in which most of the members
of our party participated. After dinner hunting-parties scoured
the ice after seals, with the result of bringing in two speci-

mens, one weighing twenty-six pounds, and the other thirty-three pounds.

Sunday, July 5. All night we steamed along slowly, but at 8 A. M. we were forced once more to stop. The day has been very disagreeable, foggy, rainy, and even snowy. We have done nothing but eat and sleep. A lazily hovering ivory-gull, which ventured within near gunshot, has been added to our collections.

Tuesday, July 7. The weather yesterday was dreary and disagreeable, but to-day it seems warmer. The snow has ceased falling, although the sky is still overcast, and the fog prevents us from seeing the horizon. At noon the sun came through the clouds for a few moments, and the fog lifted sufficiently for the captain to make an observation and find that our position was latitude 74° 51'. During the afternoon the wind died down, and an attempt was made to get through the ice; but after boring and ramming the immovable pack for nearly an hour, and gaining only a ship's length, we concluded that we were burning coal for nothing. Mr. Peary, with Gibson, Astrup, Cook, and Matt, has been busy all the afternoon sawing, marking, and fitting the lumber for our Whale Sound cottage. The curing of a large number of drake-skins, intended to be made up into undershirts for winter wear, was a part of the day's work.

Thursday, July 9. Yesterday and to-day the fog lifted sufficiently at times to permit us to see the land, about forty miles distant. A good observation places us in latitude 74°

51′, and longitude about 60° W. Mr. Peary fixed the points with his pocket sextant and the ship's compass, and then made a sketch of the headlands. The ice looks rotten, but yet it holds together too firmly to permit us to force a passage.

We measured some of the floes, and found the thickest to be two and a half feet. It has seemed very raw to-day, owing largely to a slight north-west wind; and for the first time the average temperature has been below the freezing-point, being 31½° F.

Friday, July 10. This morning the rigging was covered with hoar-frost, making the "Kite" look like a "phantom ship." The fog hung heavily about us, shutting out the land completely. In the forenoon a sounding was made, but

Sailing Through the Pack.

no bottom was found at 343 fathoms. While we were at dinner, without any warning the "Kite" began to move, steam was immediately gotten up, and for an hour and a half we cut our way through the ice, which had become very rotten, large

2*

floes splitting into several pieces as soon as they were struck
by the "Kite." We made about three knots, when we
were again obliged to halt on account of a lowering fog.
Our little move was made just in time to keep up the courage of some of the West Greenland party, who were beginning to believe that we should be nipped and kept here for
the winter.

Although we realized that we were still ice-bound in the
great and much-dreaded Melville Bay pack, we could not but
enjoy at times the peculiar features of our forced imprisonment. Efforts to escape, with full promise of success, followed
by a condition of impotency and absolute relaxation, would
alternately elevate and depress our spirits to the extent of
casting joy and gloom into the little household. The novelty
of the situation, however, helped greatly to keep up a good
feeling, and all despondency was immediately dispelled by the
sound of the order to "fire up," and the dull rumbling of the
bell-metal propeller. We never tired of watching our little
craft cut her way through the unbroken pans of ice. The
great masses of ice were thrust aside very readily; sometimes
a piece was split from a large floe and wedged under a still
larger one, pushing this out of the way, the commotion caus-
ing the ice in the immediate vicinity fairly to boil. Then we
would run against an unusually hard, solid floe that would not
move when the "Kite" struck it, but let her ride right up on
it and then allow her gradually to slide off and along the edge
until she struck a weak place, when the floe would be shivered

just as a sheet of glass is shivered when struck a sharp, hard blow. The pieces were hurled against and on top of other pieces, crashing and splashing about until it seemed as though the ice must be as thick again as it was before the break-up; but the good old " Kite" pushed them aside, leaving them in the distance groaning and creaking at having been disturbed. The day has been pleasant, in spite of an average temperature of 27½°.

Tuesday, July 14. How different everything looks to us since I last wrote in this journal! Saturday the weather was, as usual, cold and foggy; and when, at 5.30 P. M., we found ourselves suddenly moving, every one was elated, hoping we would be able to get into the clear water ahead, which the mate said could be seen from the crow's-nest. Mr. Peary was particularly pleased, as he said we should then reach Whale Sound by July 15, the limit he had set for getting there. After supper he and I bundled up and went on deck, and watched the "Kite" cut through the rotten ice like butter. We had been on the bridge for some time, when Mr. Peary left me to warm his feet in the cabin. Coming on deck again, he stepped for a moment behind the wheel-house, and immediately after, I saw the wheel torn from the grasp of the two helmsmen, whirling around so rapidly that the spokes could not be seen. One of the men was thrown completely over it, but on recovering himself he stepped quickly behind the house, and I instantly realized that something must have happened to my husband. How I got to him I do not know, but

I reached him before any one else, and found him standing on one foot looking pale as death. "Don't be frightened, dearest; I have hurt my leg," was all he said. Mr. Gibson and Dr. Sharp helped, or rather carried, him down into the cabin and laid him on the table. He was ice-cold, and while I covered him with blankets, our physicians gave him whisky, cut off his boot, and cut open his trousers. They found that both bones of the right leg had been fractured between the knee and the ankle. The leg was put into a box and padded with cotton. The fracture being what the doctors pronounced a "good one," it was not necessary to have the bones pulled into place. Poor Bert suffered agonies in spite of the fact that the doctors handled him as tenderly as they could. We found it impossible to get him into our state-room, so a bed was improvised across the upper end of the cabin, and there my poor sufferer lies. He is as good and patient as it is possible to be under the circumstances. The accident happened in this way. The "Kite" had been for some time pounding, or, as the whalers say, "butting," a passage through the ice, slowly but steadily forging a way through the spongy sheets which had already for upward of a week imprisoned her. To gain strength for every assault it was necessary constantly to reverse, and it was during one of these evolutions, when going astern, that a detached cake of ice struck the rudder, crowding the iron tiller against the wheel-house where Mr. Peary was standing, and against his leg, which it held pinned long enough for him to hear it snap.

Wednesday, July 15. Mr. Peary passed a fairly comfortable night, and had a good sleep without morphine to-day, consequently he feels better. As for myself, I could not keep up any longer, and at 11 A. M., after Dr. Cook had dressed the leg and made an additional splint, I lay down, and neither moved nor heard a sound until after five o'clock. This was the first sleep I have had since Friday night. Dr. Cook, who has been more than attentive, has made a pair of crutches for the poor sufferer, but he will not be able to use them for a month.

We find to-day that our latitude is 75° 1′, and our longitude 60° 9′; consequently our headway has been very slow. It seems as if when the ice is loose the fog is too thick for us to travel in safety, and when the fog lifts the ice closes in around us. Once to-day the ice suddenly opened and a crack which visibly widened allowed us to make nearly four miles in one stretch. Throughout much of the night and day we steamed back and forth and hither and thither, trying to get through or around the ice, and to prevent the "Kite" from getting nipped between two floes. A little after supper the fog suddenly closed in upon us, and before we could complete the passage of a narrow and tortuous lead, through which we were seeking escape from the advancing floes in our rear, we were caught fast between two large pans. The ice was only about fourteen inches thick, and there was but little danger of the "Kite" being crushed; still, Captain Pike, with the memories of former disasters fresh in his mind, did not relish the situation, and blasted our way out with gunpowder at 8.15 P. M. This is our first "nip."

An hour later the captain called down to me to come up at once, as a bear was advancing toward the ship. The boys had been watching and longing for a bear ever since we left New-York, and many false alarms had been given; but here was a real live polar coming straight for the "Kite." A very, very pretty sight he was, with black snout, black eyes, and black toes. Against the white snow and ice, he seemed to be of a cream color. His head was thrown up as he loped along toward us, and when, within a short distance of the "Kite," a gull flew over his head, he made a playful jump at it, all unconscious of the doom which awaited him. Eleven men with guns were stooping down on the quarter-deck waiting for the captain to give the word to fire. A bullet disabled one of the fore legs, while another struck the animal in the head, instantly

Bruin at Rest.

dyeing it crimson; the bear stopped short, wheeled round, fell over on his head, and then got up. By this time it was simply raining bullets about the poor beast; still he staggered on toward the water. Gibson, who had jumped on the ice as soon as he fired, was now close to him, and, just as he started to swim away, put a ball in his neck, which stopped him short. A boat was lowered, and he was brought alongside the "Kite." He measured seven feet one inch in length, and we estimated his weight at from eight to ten hundred pounds.

Friday, July 17. Last night was the worst night my poor husband has had. His leg pained him more than it had done so far, and he begged me to give him a sedative, which, with the doctor's consent, I did; but even then his sleep was disturbed to such an extent that it amounted to delirium. He would plead with me to do something for his leg. After doing everything I could think of, I said, " Can't you tell me where it hurts you most, and what you think might help you ? " His answer was, " Oh, my dear, pack it in ice until some one can shoot it ! " In this way he spent the night, and this morning he was thoroughly exhausted. Dr. Cook has succeeded in making his leg more comfortable, and now he sleeps. It seems very hard that I cannot take him away to some place where he can rest in peace.

Tuesday, July 21. Since last writing in my journal, four days ago, we have been steadily nearing Cape York, and we hope soon to clear the ice of Melville Bay, and pass into the open North Water beyond. Our hopes have, however, so often been disappointed that day by day, even when in full view of the land, we become less and less confident of ever being able to disengage ourselves from our confinement. Huge grounded bergs still hold the ice together, and until they show signs of moving there is little prospect of a general break-up.

On Saturday a bear with two cubs was seen on the ice ahead of us, and immediately every man was over the side of the vessel making for the animals. The mother, with a tender affection for her young, guided an immediate retreat, herself taking the rear, and alternately inciting the one cub and then the other to more rapid movement. Our boys were wholly unacquainted with the art of rapid traveling on the rough and hummocky ice, and before long the race was admitted to be a very unequal one; they were all quickly distanced. One of the men, in the excitement of the moment, joined in the chase without his gun, and, even without this implement, when he returned to the "Kite" he was so out of breath that he had to be hauled up the sides of the vessel like a dead seal. He lay sprawling and breathless on the deck for at least five minutes, much to the merriment of the crew and the more fortunate members of the party. A round weight of over two hundred pounds was responsible for his discomfiture. Monday morning about two o'clock the fog suddenly lifted, and we found ourselves almost upon the land. The visible shore extended from Cape York to Wolstenholme Island, and we could clearly distinguish Capes Dudley Diggs and Atholl. I held a looking-glass over the open skylight in such a way that Mr. Peary could see something of the outline of the coast. Poor fellow! he wanted to go on deck so badly, thinking that if he were strapped to a board he could be moved in safety, but the doctor persuaded him to give up the thought. As the doctors have all agreed that in six months his leg will be as good as it ever was, he refuses to consider the idea of returning on the "Kite"; as for myself, now that we have started, I want to keep on too. The air is almost black with flocks of the little auk, and a party on the ice to-day brought in sixteen birds in a very short time.

Wednesday, July 22. Drs. Hughes and Sharp brought in sixty-four birds as the result of an all-night catch. We are still in the ice, with no signs of our getting out, although the captain says we have drifted twenty miles to the northward since Monday morning. We are now abreast of Conical Rock. Second Mate Dunphy has just reported seeing from the crow's-nest a steamer off Cape York, but it is not visible to the naked eye, and we are in doubt as to what it is.

Friday, July 24. The steamer did not materialize; either the mate was mistaken or the vessel drifted away from us. The ice parted early yesterday morning, much to everybody's relief, and we have since been pushing steadily on our course. The long line of table-topped bergs off Cape York, some of which measured not less than two hundred to three hundred feet in height, and perhaps considerably over a mile in length, is visibly moving over to the American waters, and to this disrupting force we are doubtless largely indebted for our liberation. The scenery of this portion of the Greenland coast is surpassingly fine. The steep red-brown cliffs are frequently interrupted by small glaciers reaching down to the water's edge. The entrance to Wolstenholme Sound, guarded as it was by huge bergs, was particularly beautiful. Saunders

Island in the distance, and Dalrymple Rock immediately in the foreground, stood up like great black giants, contrasting with the snow-white bergs surrounding them and the red cliffs of the mainland on either side. Whenever anything particularly striking or beautiful appears, I am called on deck, and with my hand-glass placed at the open transom over Mr. Peary's head, manage to give him a faint glimpse of our surroundings. At nine o'clock this evening we rounded Cape Parry, and about ten o'clock stopped at the little Eskimo village of Netchiolumy in Barden Bay, where we hoped to obtain a native house, sledge, kayak, and various native utensils and implements for the World's Columbian Exposition. Our search-party found only three houses in the settlement, and the lonely inhabitants numbered six adults and five children; five dogs added life to the solitude. These people had quantities of sealskins and narwhal tusks, many of which were obtained in exchange for knives, saws, files, and tools in general. Wood of any kind, to be used in the construction of sledges, kayak frames, and spear- and harpoon-shafts, was especially in demand; they cared nothing for our woven clothing, nor for articles of simple show and finery. We stopped this morning at Herbert Island, where we had hoped to visit a native graveyard, but no graves were found.

FRAGMENTS FROM THE HEILPRIN GLACIER—HEAD OF INGLEFIELD GULF.

CHAPTER III

ESTABLISHING OURSELVES

Arrival at McCormick Bay — Selecting the Site for the House — Temporary Quarters — Hurrying the Erection of the House — White Whales — Departure of the "Kite"— Alone on the Arctic Shore — A Summer Storm — Arctic Picnicking — The First Birthday and the First Deer — Birthday-dinner Menu — Departure of the Boat Party for Hakluyt and Northumberland Islands after Birds and Eskimos — Occupations during their Absence — Return of the Party with an Eskimo Family.

Sunday, July 26. Mr. Peary is getting along nicely. His nights are fairly comfortable, and consequently he feels much better by day; his back now troubles him more than his leg. Yesterday morning at three o'clock he was awakened and told that the ice prevented our getting to Cape Acland, and that we were just abreast of McCormick Bay, and could not proceed further into the sound. He accordingly decided to put up our quarters on the shores of this bay. It was now a question as to which side of the bay would be most favorable for a home. At 9 A. M., together with several members of our party, I rowed over to the southeast shore, and walked along the coast for about three miles, prospecting for a site, and made a provisional choice of what seemed a desirable knoll. We returned to the "Kite" about noon. After dinner Professor Heilprin, Dr. Cook, Astrup, and three others went over to

31

the other shore, and toward evening they returned with the
report that the place was perfectly desolate and not at all
suitable for a camp. After supper we returned to the southeast
shore to see if we could improve on the location selected in the
morning, but after tramping for miles came back to the old site.
While it cannot in truth be said that the spot is a specially
attractive one, it would be equally untrue to describe it as
being entirely devoid of charm or attraction.   Flowers bloom
in abundance on all sides, and their varied colors,—white,
pink, and yellow,—scattered through a somewhat somber
base of green, picture a carpet of almost surpassing beauty.
Rugged cliffs of sandstone, some sixteen hundred to eighteen
hundred feet high, in which the volcanic forces have built up
long black walls of basalt, rise steeply behind us, and over
their tops the eternal ice-cap is plainly visible.   Only a few
paces from the base of the knoll are the silent and still partially ice-covered waters of the bay, which extends five miles
or more over to the opposite shore, and perhaps three times
that distance eastward to its termination.   A number of
lazy icebergs still stand guard between us and the open
waters of the western horizon, where the gray and ice-
flecked bluffs of Northumberland and Hakluyt Islands disappear from sight.

This morning the members of our party went ashore with
pickaxes and shovels, and they are now digging the foundations of our " cottage by the sea."   They are also putting up
a tent for our disabled commander, whence he can super-

ON THE BEACH OF McCORMICK BAY.

intend the erection of the structure. The men are working in their undershirts and trousers, and it is quite warm enough for me to stay on deck without a wrap, even when I am not exercising; yet, if we had this temperature at home, we should consider it decidedly cool. I have had oil-stoves taken ashore for the purpose of heating the tent in case it becomes necessary.

Wednesday, July 29. The last three days have been busy ones for me, being obliged to attend to all the packing and

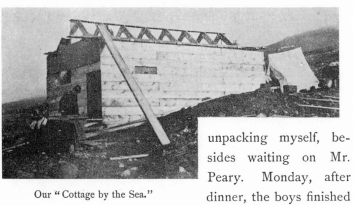

unpacking myself, besides waiting on Mr. Peary. Monday, after dinner, the boys finished

Our " Cottage by the Sea."

digging the foundations. Mr. Peary was then strapped to a board, and four men carried him from the " Kite " into a boat. After crossing the bay he was carried up to the tent just back of where the house is being erected, and placed on a rough couch. He is near enough to superintend the work, and everything is progressing favorably.

Last night was a queer one for me. All the boys slept on board the " Kite," leaving me entirely alone with my crippled

3

husband in the little shelter-tent on the south shore of
McCormick Bay. I had forgotten to have my rifle brought
ashore, and I could not help thinking what would be the best
thing for me to do in case an unwelcome visitor in the shape
of a bear should take it into his head to poke his nose into
the tent. While I was lying awake, imagining all sorts of
things, I heard most peculiar grunts and snorts coming from
the direction of the beach, and on looking out saw a school of
white whale playing in the water just in front of our tent.
They seemed to be playing tag, chasing each other and diving
and splashing just like children in the water. I was surprised
at their graceful movements as they glided along, almost
coming up on the beach at times. The night passed unevent-
fully, but I decided to have Matt sleep on shore to-night,
should the others go on board the "Kite" again. In case of
a sudden wind-storm I could not steady the tent alone, and
some one ought to be within calling distance.

As the members of the returning party come to bid us
good-by it makes me feel very, very homesick; but a year
will soon pass, and then we too shall return home. The professor has kindly offered to see mama, and do for her what
he can in the way of keeping her posted.

Early Thursday morning, July 30, those of our party who
had slept aboard ship—that is, all except Mr. Peary, Matt, and
myself—were aroused and told they must "pull for the
shore," as the "Kite" was going to turn her nose toward

home. Not being accustomed to the duties of housekeeper and nurse, I was so completely tired out that I slept soundly and knew nothing of the cheers and farewell salutes which passed between the little party who were to remain in the far North, and those on board the "Kite," who would bring our friends the only tidings of us until our return in '92. Mr. Peary remarked on the cheerfulness of our men. Less than five minutes after the boat grated on the beach he heard the sound of the hammer and the whistling of the boys.

Three or four hours after the "Kite" left McCormick Bay a furious wind and rain storm swept down upon us from the cliffs back of our house. The boys continued the work on the roof as long as possible, hoping to be able to get the whole house under cover, but the fury of the storm was such as to make it impossible for them to keep their foothold on the rafters, and they were obliged to seek shelter under what there was of the roof. At meal-time they all crowded in our little 7 x 10 canvas tent, sitting on boxes and buckets, and holding their mess-pans in their laps. These I supplied with baked beans, stewed corn, stewed tomatoes, and corned beef, from the respective pots in which they had been prepared. The rain dashed against the tent, and the wind rocked it to and fro. Every little while one of the guy-ropes would snap with a sound like the report of a pistol, and one of the boys would have to put his dinner on the ground and go out into the storm and refasten it, for these ropes were all that kept our little tent from collapsing. The meal completed, the boys returned to the house, where they had more room, even if they were not more comfortable.

I never shall forget this wretched night following the departure of the "Kite." The stream which rushed down the sides of the cliffs divided just back of the tent, and one arm of it went round while the other came through our little shelter. The water came with such force that in a few moments it had made a furrow down the middle of the tent floor several inches deep and nearly the entire width of the floor space, through which it rushed and roared. All night long I was perched tailor-fashion on some boxes, expecting every moment to see the tent torn from its fastenings and the disabled man lying by my side exposed to the fury of the storm. Our only comfort, and one for which we were duly thankful, was that during this "night" of storm we had constant daylight; in other words, it was just as light at two o'clock in the morning as it was at two o'clock in the afternoon. When it was time for breakfast, I lighted the oil-stove, which I had fished out of the water just as it was about to float away, and made some coffee, and we breakfasted on coffee, biscuit, and corned beef.

This state of affairs continued until the afternoon, when the storm finally abated and the boys began work again on the roof. The water in the tent subsided, and by putting pieces of plank down I could again move about without sinking into the mud, and I at once set to work to get the boys a square meal.

By Saturday morning our habitation was under cover, the stove put up temporarily, with the stovepipe through one of the spaces left for a window, and a fire made from the blocks and shavings that had escaped the flood. The house was soon comparatively dry,—at least it did not seem damp when compared with the interior of the tent,—and Mr. Peary was carried in and placed on a bed composed of boxes of provisions covered with blankets. Although we had no doors or windows in place, we felt that it might rain and storm as much as it pleased, and it would not interfere with finishing up the house and getting the meals, two very important items for us just then.

Gradually our home began to have a finished appearance: the inside sheathing was put on, and the doors and windows put in place. We had no more violent wind-storms, but it rained every day for over a week. At last, on August 8, there was no rain; and, as it was Matt's birthday, Mr. Peary told the boys after lunch to take their rifles and bring in a deer. One of the rules of our Arctic home was that each member's birthday should be celebrated by such a dinner as he might choose from our stock of provisions. Before going out Matt chose his menu, which I was to prepare while the hunters were gone. The plum-duff, however, he mixed himself, as he had taken lessons from the cook on board the "Kite." After every one had gone, Mr. Peary surprised me by saying he intended to get up and come into the room where I was preparing the dinner. Only the day before the

3*

doctor had taken his leg out of the box and put it in splints, and he had been able for the first time since July 11 to turn on his side. I tried to persuade him to lie still for another day, but when I saw that he had set his heart on making the effort, I bandaged up the limb and helped him to dress. Then I brought him the crutches which Dr. Cook had made while we were still on board the ship, and with their aid he came slowly into the other room. Here, through the open door, he could watch the waves as they rose and fell on the beach about one hundred yards distant, while I prepared the "feast." The bill of fare that Matt selected was as follows:

Mock-turtle soup.

Stew of little auk with green peas.

Broiled breasts of eider-duck.

Boston baked beans, corn, tomatoes.

Apricot pie, plum-duff with brandy sauce.

Sliced peaches.

Coffee.

With the soup I served a cocktail made by Mr. Peary after a recipe of his own, and henceforth known by our little party as "Redcliffe House cocktail"; with the stew, two bottles of "Liebfrauenmilch"; and with the rest of the dinner, "Sauterne." About five o'clock we heard the shouts of the boys, and on going out I saw them coming down the cliffs heavily laden with some bulky objects. I rushed in and reported the facts in the case to Mr. Peary, who immediately said, "They are bringing in a deer. Oh, I must get out!" So out he hobbled, and to the corner of the house, where he had a good view of the returning hunters. As soon as he saw them he said, "Get me my kodak. Quick!" and before the boys had recovered from their surprise at seeing Mr. Peary, whom they had left confined to his bed, standing on three legs at the corner of the house, the first hunting-party sent out from Redcliffe had been immortalized by the ever-present camera. The boys were jubilant over their success, and brought back appetites that did justice to the dinner which was now nearly ready. At six o'clock we all sat down at the rude table, constructed by the boys out of the rough boards left from the house, and just large enough to accommodate our party of seven. We had not yet had time to make chairs, so boxes were substituted, and we managed very nicely. We had no table-cloth, and all our dishes were of tin, yet a merrier party never sat down to a table anywhere. Three days afterward we repeated the feasting part of the day, with a variation in the bill of fare, in honor of the third anniversary of our marriage, and this time we sampled the venison, which we found so delicious that the boys were more eager than ever to lay in a stock for the winter.

The next day, August 12, Mr. Peary sent all the boys, except Matt, in one of our whale-boats, the "Faith," to search Herbert and Northumberland Islands for an Eskimo settlement, and if possible to induce a family to move over and settle down near Redcliffe House. The man could show us the best hunting-grounds, and assist in bagging all kinds of game, while the woman could attend to making our skin boots, or kamiks, and keeping them in order.   They were also instructed to visit the loomeries, as the breeding places of the birds are called, and bring back as many birds as possible.

During their absence Matt was at work on our protection wall of stone and turf around Redcliffe, and Mr. Peary busied himself as best he could in making observations for time, taking photographs, and pressing flowers and other botanical specimens which I gathered for him.   He even ventured part of the way up the cliffs at the back of the house, but this was slow and laborious work.   The ground was so soft that his crutches would sink into it sometimes as much as two feet. The weather continued bright and balmy, and I did not feel the necessity of even a light wrap while rambling over the hills.   What I did long for was an old-fashioned sunbonnet made of some bright-colored calico, and stiffened with strips of pasteboard, for the sun was burning my face and neck very badly.   The boys returned at the end of a week, bringing with them a native man named Ikwa; his wife, Mané; and two children, both little girls—Anadore, aged two years and six months, and a baby of six months, whom we called Noyah (short for Nowyahrtlik).

CHAPTER IV

HUNTS AND EXPLORATIONS

Ikwa and his Family — Present of a Mirror — August Walrus Hunt — Preparations for Sending out the Depot Party — Departure for Head of McCormick Bay — First Herd of Reindeer — Exciting Experiences in Tooktoo Valley — Packing the Things up the Bluffs — The Inland Ice Party Off — Return to Redcliffe — A Foretaste of Winter.

These Eskimos were the queerest, dirtiest-looking individuals I had ever seen. Clad entirely in furs, they reminded me more of monkeys than of human beings. Ikwa, the man, was about five feet two or three inches in height, round as a dumpling, with a large, smooth, fat face, in which two little black eyes, a flat nose, and a large expansive mouth were almost lost. His coarse black hair was allowed to straggle in tangles over his face, ears, and neck, to his shoulders, without any attempt at arrangement or order. His body was covered with a garment made of birdskins, called by the natives "ahtee," the feathers worn next the body, and outside of this a garment made of sealskin with the fur on the outside, called "netcheh." These garments, patterned exactly alike, were made to fit to the figure, cut short at the hips, and coming to a point back and front; a close-fitting hood was sewed to the neck of each garment, and invariably pulled over his head

when he was out of doors. His legs were covered with sealskin trousers, or "nanookies," reaching just below the knee,
where they were met by the tanned sealskin boots, called by
the natives "kamiks." We learned later that sealskin trousers

were worn only by
those men who were
not fortunate enough
or able to kill a bear.
In winter these men
wear dogskin trousers,
which are as warm as
those made of bear-
skin, but not nearly so
stylish. Winter and
summer the men wear
stockings reaching to
the knee, made of the
fur of the Arctic hare.
At first I thought
the woman's dress was
identical with that of
the man, and it puz-
zled me to tell one from
the other ; but in a

Mané and Anadore.

day or two I had made out the many little differences in the
costumes. The woman, like the man, wore the ahtee and
netcheh made respectively of the birdskins and sealskin.

They differed in pattern from those of the man only in the back, where an extra width is sewed in, which forms a pouch extending the entire length of the back of the wearer, and fitting tight around the hips.   In this pouch or hood the baby is carried: its little body, covered only by a shirt reaching to the waist, made of the skin of a young blue fox, is placed against the bare back of the mother; and the head, covered by a tight-fitting skull-cap made of sealskin, is allowed to rest against the mother's shoulder.   In this way the Eskimo child is carried constantly, whether awake or asleep, and without clothing except the shirt and cap, until it can walk, which is usually at the age of two years; then it is clothed in skins, exactly as the father if it is a boy, or like the mother if a girl, and allowed to toddle about.   If it is the youngest member of the family, after it has learned to walk it still takes its place in the mother's hood whenever it is sleepy or tired, just as American mothers pick up their little toddlers and rock them.

The woman's trousers, or nanookies, are made of foxskin, and are hardly anything more than "trunks"; these are met by the long-legged boots, or kamiks, made of tanned sealskin, and the long stockings, or "allahsy," of reindeer fur.   Altogether this family appeared fully as strange to us as we did to them.   They had never before seen woven material, and could not seem to understand the texture, insisting that it was the skin of some animal in America.

They brought their dog, a sledge, a tent, a kayak (or canoe), and all their housekeeping utensils and articles of furniture, which consisted of two or three deerskins, on which the family slept; a stove made of soapstone and shaped like our dust-pans, in which they burned seal fat, using dried moss as a wick; and a dish or pot made of the same material, which they hung over their stove, and in which they melted the ice for drinking purposes and also heated their seal and walrus meat (I say heated, for we would hardly call it cooked when they take it out of the water). The skin tent put up, and these articles put in place, the house was considered furnished and ready for occupancy. Wood being almost impossible to procure, the tent was put up with narwhal tusks, which are more plentiful and answer the purpose. The tent itself is made of sealskin tanned and sewed together with narwhal sinews. These people were very curious to see the white woman, who, they were given to understand, was in the American "igloo" (house); and when Mr. Peary and I came out, they looked at both of us, and then Ikwa asked, "Soonah koonah?" Of course we did not know then what he wanted, but he soon made us understand that he wished to know which one of the two was the woman. I delighted him, and won his lasting favor, by making him a present of a knife. His wife, Mané, was almost overwhelmed by a gift of some needles; while Anadore, the elder of the two children, amused herself by making faces at her image in a small mirror that I had presented to her. It was the first time these people had seen themselves, and the parents were as much amused as the children. They asked many questions, but as we could not understand them any

A SUMMER DAY.—IKWA AND FAMILY.

more than they knew what we were talking about, the whole conversation was decidedly more amusing than instructive.

Later in the day the boys launched the whale-boat, and Mr. Peary, Gibson, Verhoeff, Matt, and myself, with our new man Ikwa, went down to Cape Cleveland, two and a half miles from Redcliffe, where the boys had beached a walrus killed by them while crossing Murchison Sound.   It was very interest-

ing to watch Ikwa cut up this enormous animal, weighing more than 1500 pounds, with an ordinary six-inch pocket-knife. So precisely did he know just where every joint was, that not once did he strike a bone, but cut the entire animal up into pieces which could be easily handled by one man, as though it had been boneless. This done, the pieces were packed in the

Ikwa and his Quarry.

boat, preparatory to taking them to Redcliffe.   Here at Cape Cleveland we found the grass very green, and in places over two feet high.   This unusual growth is explained by the presence of blubber caches, seal caches, and the ruins of an Eskimo village. We gathered many flowers, among which the yellow Arctic poppy was the most prominent, and also shot a number of little auk and a few gulls and eider-ducks. Mr. Peary hobbled along the beach on his crutches, around the cape, and had his first view up Whale Sound and Inglefield Gulf. On our return to Redcliffe, all the meat was hung up back of the house to be used in the winter for dog-food and as an occasional treat for our Eskimo family. It was a little too strong for our taste, and we decided we would resort to it only in case we were unsuccessful in getting deer.

A few days after this, early in the morning, Ikwa came running into our house, apparently much excited, crying, "Awick! Awick!" This we had learned was walrus. The boys tumbled out of their beds, and in a very few moments were in the boat with Ikwa, pulling in the direction of a spouting walrus out in McCormick Bay. In a short time they returned with a large mother walrus and her baby in tow. The mother had been killed, but the baby—a round bundle of fat about four feet long—was alive, and very much so, as we found out a little later. Mr. Peary wanted to get photographs of the little thing before it was shot, and while he was dressing, a task which was of necessity slow, the boys came into the house, leaving the baby walrus about a hundred yards up on the beach. Suddenly we heard cries for help coming from the shore. On stepping to the window, I saw one of the most comical sights I had ever seen. The little walrus was slowly but surely making his way to the waters of the bay. Mané with her baby on her back was sitting in the sand, her heels dug into it as far as she could get them, holding on to the line attached to the walrus, without apparently arresting its progress in the least, for she was being dragged through the gravel and sand quite rapidly. While I looked, Matt came rushing to her assistance, and taking hold of the line just ahead of where Mané held it, he gave it one or two turns about his wrists, and evidently thought all he had to do would be to dig his heels into the sand and hold back; but in an instant he was down in the sand too, and both he and Mané were plowing along, the sand flying, and both shouting lustily for help. So strong was this little creature that had not the other boys rushed out and secured him, he would easily have pulled Matt and Mané to the water's edge, where, of course, they would have let him go, and he would have been a free walrus once more. I have always regretted that I did not get a "kodak" of the scene.

It was now the end of August, and active preparations were in progress for sending a party with provisions to establish an advance depot on the inland ice for the spring sledge journey across the great ice desert to the northern terminus of Greenland. It was decided that Astrup, Gibson, and Verhoeff should go on this trip, while Dr. Cook and Matt remained with Mr. Peary and myself at Redcliffe.

On September 3, all arrangements having been perfected for the inland ice party, every one in the settlement, except

Matt and Mané with her children, sailed for the head of McCormick Bay, where it had been decided that the boys should ascend the cliffs and attack the ice. Redcliffe House is about fifteen miles from the head of the bay, and this distance had to be rowed, for we got no favoring breeze. It was late in the evening when we rounded a point of land whence we could see the green valley stretching from the water's edge back to the giant black cliffs, which here form the boundaries of the inland ice. The landscape was a beautiful one. As I looked I beheld moving objects on one of the hillsides, which, seen through the glass, seemed to me to be the size of a cow. We at once knew they were reindeer, and their apparent size was due to mirage. Astrup was landed with a Winchester at a point where he could go round and come upon the grazing herd from behind the hill; it was hoped they would not see him, and that he would bag quite a number. After landing Astrup we kept on until we were opposite the center of the valley; here our boat was run ashore, and we decided to camp.

Mr. Peary told me to take a run over the rocks and down the valley in order to get warm, as I had become chilled from sitting in the boat and not exercising for several hours; so after seeing him safely on the little knoll about twenty feet above the shore-line, where we intended to make camp, I strolled away. Upon climbing the hill, just back of the camping-ground, I came in sight of the herd of deer which we had seen from the boat, and as I watched them I saw the smoke

and heard the report of Astrup's rifle. In an instant they
were scampering off in every direction, and although Astrup
fired shot upon shot not one dropped. One of the animals,
however, after running some distance, stumbled and fell, lay
still for an instant, then got up, ran on a few yards, and fell
again. As it did not rise I judged it had received one of
Astrup's bullets, and forgetting how deceptive distances are
in the pure, clear air, I started on the run toward the prostrate
creature, apparently not more than a mile distant. Happen-
ing to look back, I saw Dr. Cook and Ikwa coming in my
direction, and waited for them. On reaching me the doctor
said they were on their way to help Astrup bring in his game.
I called his attention to the little white spot on the green
grass, and told him it was a deer, and that I had seen it drop.
As we could see nothing of Astrup, we decided to take care
of the animal. Dr. Cook had his rifle loaded with twelve
cartridges, Ikwa had a muzzle-loader charged, and an extra
load for it besides, and I had on my cartridge-belt and revolver (a 38-caliber Colt). After walking—or trotting would
perhaps express it better—for some distance, we came to a
stream that flowed down the center of the valley throughout
its length, which we had to cross in order to reach our destination. Fortunately the doctor had on his long-legged rubber
boots, for we soon saw that the only way to get on the other
side was to wade the stream. We tried it at different places,
and finally the doctor found a place where he could cross.
First taking his rifle and my revolver and belt of cartridges

4

over, he returned for me and carried me across; then we
continued in the direction of the white spot, which all this
time had not moved.   After traveling for nearly an hour we
were near enough to see that beside the prostrate deer stood
a tiny black-and-white creature, a fawn.   Whether it saw us
and whispered to its mother, I do not know; but immediately
after we had made out the little one, the mother deer raised
her head, looked at us, then rising slowly, started off at a
moderate walk.   We quickened our steps, and so did she.
When within three hundred yards, Dr. Cook discharged his
rifle several times, but only succeeded in wounding her in the
fore leg, which did not seem to retard her progress in the least.
Several times we were near enough to have shot her without
any trouble, but we were so excited—a case of buck-fever, I
believe the hunters call it—that she escaped every shot.   To
add to our difficulties the deer made for a neighboring lake,
and in the effort to stop her before she reached it, we fired
shot after shot until the doctor's rifle was empty.   There was
now nothing for us to do but stand around and crouch behind
the boulders in the hope that the poor wounded animal would
come ashore within pistol-shot range.   It was evident that she
was too weak to swim across, and it was very touching to see
how the little fawn would support its mother in the water.
Once or twice she tried to climb out on the ice-foot, but the
ice was not strong enough, and broke beneath her weight.   We
were thoroughly chilled and hungry by this time, but disliked
the idea of returning empty-handed to camp after such a long

absence. At last, just as we were talking of returning, we saw Astrup in the distance, and called to him to join us. When he came up to us he said he had had no luck. He had a few cartridges left in his rifle, which he expended on our victim without, however, harming her in the least. Astrup then urged us to return, as he, too, was tired out; but we were loath to leave our wounded deer, especially as we now knew it was only a matter of time when we should get her, for she could not hold out much longer. Nearer and nearer she came to the ice, finally leaning against the edge as if to gather strength, when suddenly the doctor darted over the ice-foot into the icy water, and before the startled animal realized his intention, he had her by her short horns, which were still in the velvet, and was pulling her slowly ashore. The little one then left its mother for the first time, ran as fast as it could over the rocks, and disappeared behind the cliffs.

The doctor had some trouble in pulling the wounded animal out on the ice, which kept constantly breaking. All this time he was standing knee-deep in the ice-cold water, and before long he had to call to us to relieve him, his feet and legs being so numb that he could stand it no longer. As Astrup had on low shoes, he did not feel like wading out to the doctor, who was rubbing and pounding his feet, so I went to his relief. My oil-tan boots kept the water out for some time. Although I could not drag the poor creature out on the ice, still I had no difficulty in holding her, as she made no resistance whatever. After the doctor had somewhat restored his circulation, he came to me, and together we pulled the wounded animal out. Then I was asked to kill her with my revolver, but I could not force myself to do it, and Astrup took the weapon and put her out of her misery. We placed the body on a large flat rock, piled boulders on it, and left it. Both Dr. Cook and I were thoroughly cold by this time, and we all hurried toward camp. It was now nearing midnight, and I had been away from camp since six o'clock. It was hard to realize the time of day, as the sun was shining just as brightly as in the early afternoon. We soon reached the river, and across it the poor doctor had to make three trips: first to carry the rifles over, then to come back for me, and then to go after Astrup. As this last load weighed 183 pounds, and the current was very swift, progress was of necessity slow. The doctor had to feel his way, and did not dare to lift his feet from the bottom. At last we were all safely over. Ikwa, who had taken off his kamiks and stockings and waded the stream, was lying flat on his back on a mossy bank nearly convulsed with laughter at the sight of the doctor carrying Astrup. Once across the river we redoubled our speed, and soon reached camp, where I found Mr. Peary, with Gibson and Verhoeff, anxiously awaiting me.

The next two days the boys spent in packing their provisions and equipment over the bluffs to the edge of the ice, while I stayed in camp and cooked, and Ikwa put in his time hunting. On the fourth day, Monday, September 7, right after lunch, the boys left with their last load, and in spite of the

LOOKING DOWN INTO THE SUN GLACIER FJORD FROM THE ICE-CAP.

snow, which had been falling lightly all day, determined to keep on to the inland ice. Dr. Cook accompanied them, helping them carry their provisions to the edge of the ice, and on his return we were to start for Redcliffe.

Just as everything had been stowed away in the boat, a wind-storm came down upon us which threatened to blow our little craft upon the rocks. The sea was rough and the wind cold, which made the time of waiting for the doctor seem very long. At last we were joined by our companion, who told us that he had left the inland ice party ensconced in their sleeping-bags, and that it was snowing furiously upon the ice-cap. When we reached Redcliffe seven hours later, we found everything white and about ten inches of snow on the ground.

The Crew of the " Faith."

Cook.   Ikwa.   Gibson.   Astrup.   Verhoeff.

4*

CHAPTER V

BOAT JOURNEYS AND PREPARATIONS FOR WINTER

Return to Head of McCormick Bay for Deer — Footprints on the Shore — Success-
ful Deer Hunt — Meeting with the Returning Inland Ice Party — Astrup and
Gibson Make a Second Attempt on the Ice-cap — Attempted Boat Trip up
Whale Sound — Stopped by the New Ice — Exciting Battle with Walrus —
Dr. Cook and Matt Tramp to Nowdingyah's — Last of the Boat Trips — Setting
up the Stove — My Experience with a Snow-slide — Final Return of the Inland
Ice Party — Preparing Redcliffe for Winter.

We were all pretty tired the next day, and Mr. Peary decided to wait another day or two before starting on a second
hunting-expedition to the head of the bay. It was Thursday
morning, September 10, when we nailed up our doors and, out
of regard for "social custom," tacked a card on the front door,
which read: "Have gone to Tooktoo Valley for two or three
days' hunt. Visitors will please leave their cards," and then
headed our boat eastward.

In order to avail ourselves of the breeze, we were obliged to
cross the bay and then tack. When about half-way it was
decided to run ashore and prepare lunch. As soon as the
keel of the boat grated on the sand, Ikwa jumped out to make
the bow-line fast, but he had hardly touched the ground before he gave utterance to a cry of surprise, and pointed to footprints in the sand. In a moment we were all excitement.

The footprints were those of two persons walking in the direction of Redcliffe. What a peculiar sensation it is to find signs of human beings in a place where you believe yourself and party to be the only inhabitants! After examining them carefully, Ikwa said Gibson and Verhoeff had passed down the beach that morning. This worried Mr. Peary, for the supposition was that something must have happened to one of the party, and the other two were bringing him to Redcliffe. He was reassured, however, in a few minutes; for on following the footprints a little distance, I found the prints of all three of the boys, and we knew that the inland ice party had returned. Knowing that they would make themselves comfortable at the house, Mr. Peary decided to keep on to the hunting-grounds, which we reached in the early afternoon. During our three days' stay in this lovely valley, Matt and Ikwa bagged nine deer; I myself went hunting once or twice, but without success. Most of my time was devoted to taking photographs of the glaciers in the vicinity, and keeping camp. The sand along the shore was too deep and the hills were too steep for Mr. Peary to take long walks in any direction, and he was glad to have company in camp.

On Monday we loaded our boat with the trophies of the chase, and sailed for home. When within three and a half miles of the house, we saw Astrup and Verhoeff coming up the beach, and we immediately hailed them, and pulled for the shore. They got into the boat, and during our sail home Astrup told of the continued storm on the ice-cap; how the deep snow had prevented their making more than one or one and a half miles per day ; that Verhoeff had frozen his face, and that they had then decided to return to Redcliffe, report the condition of the traveling, and see if Mr. Peary wished them to keep on. After reaching Redcliffe, Mr. Peary gave the inland ice party a few days' rest, and then sent them in the " Faith," our largest whale-boat, back to the head of McCormick Bay to bring home their equipment and place all the provisions in a cache which would be easily accessible. Gibson and Verhoeff were to put in two or three days hunting deer, while Astrup was to make a careful examination of the cliffs and glaciers to ascertain the most practicable route to the ice-cap with dogs and sledges. They returned in four days, and we immediately began work changing the equipment to make it suitable for two persons instead of three, and dried out the sleeping-bags thoroughly. Three days afterward, September 22, Astrup and Gibson again set out for the inland ice.

Wednesday, September 23. This morning at 9.30 Mr. Peary, Matt, Dr. Cook, Ikwa, and myself started in the " Mary Peary " for a trip up Inglefield Gulf. There was not a breath of air stirring, and the boys had to row from the start. Before we had gone a mile, several burgomasters flew over our heads, and we next came upon a flock of eiders, but did not get within gunshot. When just off Cape Cleveland, we caught sight of several walrus in the middle of the bay, and made for them. A number of shots were fired, and some of the animals were wounded; but as Ikwa said we should be sure to find " amissu-ar " (plenty) " awick " in the gulf, we did not wait for them to come again to the surface. After a two hours' rest we proceeded up the gulf, but were stopped by the heavy new ice, which we could almost see forming in our wake. It being certain that we could not make further progress by the boat, Mr. Peary decided to have a walrus-hunt for the purpose of obtaining ivory. We could see the walrus in every direction, and headed the boat for a cake of ice with about fifteen of the

Walrus on Ice-cake.—Off Herbert Island.

creatures asleep on it. The boys were told to pull for all they were worth until the order was given to stop. Mr. Peary then took his camera, and he became so absorbed in getting his photo just right that he forgot to give the order to stop until the boat was so near the cake of ice that before anything could be done she ran on it at least four feet, throwing her bow straight up into the air. The walrus, jumping into the water from under her, careened the boat to port until she shipped water, throwing Matt flat on his back; then with a

jerk (which proved to come from an animal Ikwa had harpooned) she was righted, and we were skimming over the
water, through the new ice, towed by the harpooned walrus.
This performance lasted at least twenty minutes, during which
time the boys kept up a constant volley at the walrus that
besieged us on every side to revenge their wounded companions.   There were at least two hundred and fifty around
us at one time, and it seemed as if it would be impossible to
keep the animals from attacking us; but by steady firing
we managed to hold them at oar's length.   This kept me
busy reloading the rifles.   I thought it about an even chance
whether I would be shot or drowned.

I cannot describe my feelings when these monsters surrounded us, their great tusks almost touching the boat, and
the bullets whistling about my ears in every direction.   Whenever a volley of shots greeted them, the whole bunch jumped
into the air and then plunged under water, leaving us in doubt
as to where they would reappear.   If they should happen to
come up under the boat, we should probably be the ones to
take the plunge; this uncertainty was very exciting, especially
as the brutes went down and came up in bunches, leaving us
seventy-five or a hundred to fight while the rest plunged.

Ikwa had evidently never seen so many " awick " at one
time, and became very much frightened, finally pounding the
sides of the boat with his harpoon and yelling at the top of
his voice, in which he was joined by Matt.   When we finally
got out of the turmoil we had four heads with tusks, and would

have had more, but the bodies sank before we could secure them. As we could not proceed up the gulf in the boat, we camped about three miles southeast of Cape Cleveland. The boat was pulled up on a bit of sandy beach, and with the aid of the boat-hooks and a couple of tarpaulins we fixed up a very comfortable boat-tent.

Thursday, September 24. It was decided last night that Matt and Dr. Cook should set out on foot for "Nowding-yah's," an Eskimo camp of which we had been in search; so we had coffee early, and by eight o'clock the boys started off with their rifles and some pemmican.[1] About ten o'clock the boys came in woefully tired, vowing that they had walked forty miles, and reported finding Nowdingyah's camp, but all four igloos were deserted. Ikwa said that their owners were "pehter-ang-ito" (far away) hunting; these northern Eskimos are in the habit of leaving their settlements, to which they periodically return.

Friday, September 25. Just before we left camp at eleven o'clock, an amusing incident occurred. Ikwa, who had been skirmishing for the past hour, returned in a jubilant frame of

---

[1] It may be of interest to my readers to know just what pemmican is. The best lean beef is cut in strips and dried until it can be pulverized, then it is mixed with an equal quantity of beef suet. To this mixture are added sugar and currants to suit the taste, and the whole is heated through until the suet has melted and mixed with the other ingredients, when it is poured into cans and hermetically sealed. It is only a modification of the old-fashioned way of preserving meat when whole families drove out on the prairies and hunted buffalo. As soon as shot the buffalo was skinned and the green skin sewed into a bag, into which the meat, after it had been sun-dried and mixed with the suet, was packed. As the skin dried and shrunk, it compressed the meat, which in this way was preserved indefinitely. Pemmican is not at all unpleasant to the taste, especially if eaten with cranberry jam.

mind, and announced his discovery of a cached seal. He asked Mr. Peary if he might bring the seal to Redcliffe in the boat, saying it was the finest kind of eating for himself and family. We could not understand why this particular seal should be so much nicer than those he had at Redcliffe; but as he seemed very eager to have it, we gave him the desired permission, and off he started, saying that he would be back very soon. About half an hour later the air became filled with the most horrible stench it has ever been my misfortune to endure, and it grew worse and worse until at last we were forced to make an investigation. Going to the corner of the cliff, we came upon the Eskimo carrying upon his back an immense seal, which had every appearance of having been buried at least two years. Great fat maggots dropped from it at every step that Ikwa made, and the odor was really terrible. Mr. Peary told him that it was out of the question to put that thing in the boat; and, indeed, it was doubtful if we would not be obliged to hang the man himself overboard in order to disinfect and purify him. But this child of nature did not see the point, and was very angry at being obliged to leave his treasure. After he was through pouting, he told us that the more decayed the seal the finer the eating, and he could not understand why we should object. He thought the odor "pe-uh-di-och-soah" (very good).

At noon we passed Cape Cleveland, homeward bound, and an hour later reached Redcliffe. The house seemed very cold and chilly after the bright sunshine. Verhoeff, who had been left in charge, greeted us, and we soon had all the oil-stoves going, bread baking, rice cooking, beans heating, venison broiling, and coffee dripping, and at two o'clock all sat down to dinner and then turned in.

Tuesday, September 29. The last three days have been spent in hunting-explorations on the north shore and in preparations for the winter. The stove has been put up, the windows doubled, and the house made generally air-tight. We find the ice in the bay becoming firmer day by day, and in one of our expeditions we found it all but impossible to force the boat through it. Mr. Peary has now left off his splints and bandages, and has even laid aside his crutches. After lunch to-day I started out with a couple of fox-traps, and put them in the gorge about a mile back of the house. The day was fine, and I enjoyed my walk, although I came in for an unpleasant scare. After leaving the traps, I thought I would go over the mountains into the valley beyond, and see if I could find deer. Half-way up, about a thousand feet above sea-level, the snow began to slide under me, taking the shales of sandstone along with it, and of course I went too, down, down, trying to stop myself by digging my heels into the snow and attempting to grasp the stones as they flew by; but I kept on, and a cliff about two hundred feet from the bottom, over which I would surely be hurled if I did not succeed in stopping myself, was the only thing which I could see that could arrest my progress. At last I stopped about half-way down. What saved me I do not know. At first I was afraid to move for fear I should begin sliding again; but as I grew more courageous I looked about me, and finally on hands and knees I succeeded in getting on firm ground. I did not continue my climb, but returned to the house in a round-about way.

Mr. Peary had the fire started in the big stove, and finds that it works admirably. The trouble will be to keep the fire low enough. Ikwa indulged in a regular war-dance at the sight of the blaze, never before having seen so much fire, and for the first few moments kept putting his fingers on the stove to see how warm it was. He soon found it too hot. He has been getting his sledge, dog-harness, spears, etc., in readiness for the winter's hunt after seal.

Wednesday, September 30. Toward noon Matt came running in shouting, "Here are the boys, sir!" and sure enough Astrup and Gibson were here, bringing nothing but their snow-shoes with them. They were on the ice just a week, and estimate the distance traveled inland at thirty miles, and the greatest elevation reached at 4600 feet. They returned because it was too cold and the snow too deep for traveling. At the same time, they admit that they were not cold while on the march, and they do not think the temperature was more than 10° below zero; but as Gibson stepped on and broke the thermometer on the third day, up to which time the lowest had been −2°, they had no way of telling for certain. Gibson's feet were blistered, he having forgotten to put excelsior or grass in his kamiks. He believes that with the moral support of a large party they can easily make from ten to fifteen miles per day.

Thursday, October 1. The day has been fine; the house is gradually assuming a cozy as well as comfortable appearance under Mr. Peary's supervision. He is about from morning until night, limping a great deal, but he has put aside his crutches for good. At night his foot and leg are swollen very much, but after the night's rest look better, although far from normal. Ikwa went out on the ice to-day for some distance to test its strength. I took my daily walk to the fox-traps, and as usual found no foxes had been near them.

Sunday, October 4. Nothing of any consequence has taken place since the return of the explorers. The boys have been at work on the house, hanging blankets, putting up shelves, etc. Friday I found one of my traps sprung, and a great many tracks around it, but no fox. On Saturday we went down to the point one quarter of a mile below the house, Mr. Peary walking without cane or crutch, and set a fox-trap on the rocks near some tracks. All this time the weather has been perfect. To-day Dr. Cook tried going out on the ice, but it did not hold him. The bunks of the boys have been placed against the east side of the large room and separate curtains furnished. The winter routine of four-hourly watches throughout the twenty-four hours was begun to-day, the boys taking them in turn.

Monday, October 5. It has been cloudy all day long, but with a temperature of about 12°. It still seems warm, as there is no wind whatever. I went to my fox-traps this fore-noon, and found the view from the heights very fine. The clouds hung low, and gave a soft gray background for the blue bergs which gleamed on every side of a long black strip of water—the open sea—in the far distance. The light that fell on Northumberland Island decked it in a bright yellow, while the cliffs across the bay were black in the dark shadow.

The boys brought the "Mary Peary" up and turned her over, supporting her on pillars built of blocks of ice. Here Mr. Peary intends to put such provisions as we may need for our boat-journey home next summer, covering the whole thing with snow. The "Faith" has been turned over against the front wall, and a place fixed under her for the Newfoundland dogs, Jack and Frank. As soon as we have enough snow the house, too, will be banked in with it.

CHAPTER VI

WINTER UPON US

McCormick Bay Frozen over — First Sledge Trip to the Head of the Bay for Deer —
Shaky New Ice — First Aurora — The Strange Light on the Opposite Shore
— First Visit from the Natives — Return of our Hunting-party with Ten Deer —
More Natives — Second Severe Snow-storm of the Season — Still more Native
Visitors — Great Amusement over the White Woman — Farewell to the Sun.

Tuesday, October 6. McCormick Bay is frozen over so as
to support the dogs and sledge, and Ikwa has been on several
seal-hunts. He finds one of the holes in the ice which the
seals keep open all the winter and where they come to breathe.
Here he takes up his position, being careful not to make the
least noise. Sometimes he waits for hours before the seal
comes up, and sometimes the seal skips that hole entirely.
When it comes he drives his spear through the hole quick as
a flash into the head of the animal. In this way all the seals
are caught during the fall and winter. Ikwa went out on his
sledge with his "mikkie" (dog) after "pussy" (seal) to-day,
but did not get any.

The day has been, like yesterday, dark and cloudy, but the
temperature has been higher, averaging 20° instead of 12°;
the wind has been blowing quite fresh from the east. Mr.
Peary has set the boys at work building a sledge for a prospec-

tive journey to the head of the bay, and I have been busy all
day getting our room, or rather our bed, in order. All the
boxes have been removed from under the bed, to my great
delight, and put into the lean-to at the south end of the
house. It felt and smelt like a damp cellar under there, but
now that the air has a chance to circulate freely, I think it
will be better.

I have not been out of the house to-day. It is quite dark
at six o'clock, and on a cloudy day, as to-day, we lighted the
lamp at five o'clock.

Matt has started in as lunch-maker; this gives me nearly
all day to myself. Our first table-cloth, of unbleached cotton, also made its *début ;* it is a great improvement on bare
boards.

Wednesday, October 7. This morning, at about ten
o'clock, we started out on our first sledging-trip up the bay
in search of "tooktoo" (reindeer).

Astrup, Gibson, and Matt pulled our sledge, while Jack
and Frank, our Newfoundland dogs, and Mikkie, were har-
nessed to Ikwa's. We were delighted to see that our dogs
would pull, but Ikwa soon decided that Frank was "peeuk
nahmee" (no good), so the boys put him to their sledge, but
he preferred pulling backward to pulling forward ; by coaxing
they persuaded him to help them somewhat, but it was always
hard work to get him started after a stop.

After journeying about four miles, our Eskimo suddenly
stopped his sledge and explained that he did not want any

more deerskins, but needed "pussy" skins for his kamiks, or
boots, kayak, tupic (tent), etc., and he would leave us and
watch the seal-holes, walking home at night. He told us
how to fasten his mikkie, and then, after I had kodaked him
sitting on his seal chair at a hole, we went on. I ran along
at the upstanders of Mr. Peary's sledge, he being all alone;
but the ice being rather slippery and the dogs traveling along
at a run, I soon found it difficult to keep on my feet, and so
jumped on the sledge with Mr. Peary, and rode the greater
part of the time. The two dogs pulled us easily, the sledge
and load weighing about five hundred pounds. The dogs are
fastened to the sledge by single traces, and are guided without reins by the driver with a long whip and much shouting.
The mikkie not understanding our language, and Mr. Peary
not knowing the Eskimo terms, and not understanding the language of the whip, we had no means of guiding our team;
besides, in many places the ice had to be tested by a member
of the party going ahead with an alpenstock and "feeling"
it. Often detours had to be made, and several times we had
to rush over places where the ice buckled under us, and it
seemed as though it must let us through; for these reasons
we allowed the other sledge to take the lead. This we could
do only by stopping and letting the boys get one fourth or
one half of a mile ahead; then, giving our dogs the word, they
would scud along at the top of their speed, not making any
attempt to stop until they had caught up to the other sledge,
which they did in a few minutes. In this way we finally

reached the head of the bay shortly after six. We immediately set about putting up the tent and arranging our sleeping gear, and Mr. Peary got the stove ready and put on ice
for tea, and also a can of beans to heat. I was disabled by a
sick-headache.

During the next few days the boys made a number of unsuccessful hunting-expeditions, and their failure decided us to
return to Redcliffe. The mercury had already descended at
nights to $-4°$, yet I did not feel the low temperature, and
indeed had not felt uncomfortably cold for more than a few
minutes at a time. On the 9th, at noon, just half the disk of
the sun appeared over the top of the mountain back of the
glacier, and it was evident that we were in the shadow of the
Arctic winter. Two days later we saw the first aurora—not a
good one, however.

Monday, October 12. Back again at Redcliffe. In the
evening Matt came in very much excited, saying that there
was a moving light on the opposite shore. We all rushed
out to see it. How queer it seems to be the only human
beings on this coast! Ikwa said Eskimos were eating their
supper, and would be here to-morrow. Astrup fired a rifle.

Tuesday, October 13. About three o'clock this afternoon
Mané came in and said "Innuit" (Eskimo) was coming with
"kamutee" (sledge) and "mikkie" (dog). We ran out, and
with the aid of the glass saw two Eskimos, one of them Ikwa,
and a sledge drawn by three dogs. The strange "husky"

MY CROSS-MATCHED TEAM.—McCORMICK BAY.

turned out to be Nowdingyah, whose deserted camp we visited last month. He is much larger in every way than Ikwa, and seems bright and intelligent. When offered a knife in exchange for one of his dogs, he said the dog we wanted was the leader of his team of bear-dogs, specially trained, but he would come again by and by and then give us three others. We have now little difficulty in understanding the natives, or making ourselves understood by signs.

Saturday, October 17. The weather still continues lovely, although the days are rapidly getting shorter. Late Thursday night Ikwa, who had departed with our visitor, returned, telling us that the natives where Nowdingyah lived would soon come over to see us ; he also said that Nowdingyah had seven puppy-dogs, and this is why he was so willing to give us three. Ikwa has been laying in a supply of sealskins for a tupic and kayak, and says he will need fifteen for these articles alone ; he will require an additional supply for kamiks for himself and family. The seal is evidently the most valuable animal of the chase to the natives, who utilize every particle of it for food or clothing. About three o'clock we discovered the boys, who had gone to Five-Glacier Valley, on the opposite side of the bay, coming across the ice, and about an hour later they arrived jubilant with a load of ten deerskins, one blue fox, and one Arctic hare. Gibson had also shot two seals, which they could not, however, bring with them, as the ice was too thin for the hunters to reach their booty. Still later Ikwa came in, and said " Innuits pingersut" (Eski-

5*

mos three), "kamutee martluk" (sledges two), were coming; and in a few minutes Nowdingyah, Arrotochsuah, and Kayunah landed with two sledges and five dogs. Arrotochsuah

is an old man with gray hair, but looks exactly like a woman; Kayunah is a young man, stutters badly, and while he has a decidedly idiotic appearance he has a fox-like expression about the eyes and nose, and accordingly he has been dubbed

Arrotochsuah Fashioning a Spear.

the "Fox." Nowdingyah is the only one of the Eskimos who has hair on his face, and he has a little mustache and imperial which give to him something of a Japanese touch.

Sunday, October 18. Mr. Peary has been on the jump all day, getting odds and ends to trade with the natives. He has secured three very fine seal-spears, one walrus-lance — all with fine lines of walrus-hide — an "ikkimer" (soapstone blubber lamp), a drill, and two dogs and a sledge. The natives left early in the afternoon, the old man being tired, having been obliged to sleep out on the beach on his sledge, with no shelter, as there was no room in Ikwa's igloo; he walked about the greater part of the night to keep warm.

Monday, October 19. Astrup and Verhoeff went to-day to Cape Cleveland, and put up a flag-pole and signal for use in surveying. Mr. Peary is fixing up my lockers with cardboard, preparatory to putting up the curtains. So far the weather has been fine; we have full moon, and this makes it seem less like night, but at 8 A. M. it is still quite dark. From about eleven until two, the coloring on land, ice, snow, and sky is beautiful, all the delicate shades being brought out to best advantage. We took two short strolls, fixed up the curtains about the range and lockers, and then I did a little sewing. To-night the wind is blowing fiercely from the south.

Wednesday, October 21. Last night we had our first wind-storm since the second night of our encampment here, when I was in the tent alone with Mr. Peary, who was strapped down to a plank. The wind rattled things in a lively manner, and the boys on duty had to go out every fifteen minutes and inspect the premises to see that nothing was loosened or blown away. This wind from the southeast continued until five o'clock this morning, when it abated somewhat. The day has been cloudy. The boys have put up a snow hut for the dogs, and one for their own convenience, in which to experiment with their fur clothing and sleeping-bags.

Thursday, October 22. My brother Henry's birthday. We drank his health and prosperity in a bottle of Haute Sauterne, as we did my brother Emil's eleven days before.

My husband and I are keeping house alone. All the boys have gone on a deer-hunting expedition, while Ikwa, with the dogs, is after hares. We have had Mané here all day at work on a pattern deerskin stocking. The day has been dark and cloudy, and it has snowed lightly.

Friday, October 23. Last night it snowed a very little, and this morning it is cloudy and gloomy. We sat up till midnight, then the alarm was set for two o'clock, at which time coal had to be put on the fire—an operation to be repeated at four, and again at six. Mané has been with us all day, with her two piccaninnies, at work on deerskin stockings. The elder child, Anadore, is just at the age (two years) when she is into everything, and she tried our patience to the limit. We cannot allow Mané to take the furs to her igloo to sew, as they would be filled with "koomakshuey" (parasites), and some one must stay in the room with her to superintend her work. I am doing very little besides getting the meals and fixing up odd jobs about the rooms; reading Greely's work is about the extent of my labor. To-night at nine o'clock the thermometer is 10°, and the moon is shining brightly.

Sunday, October 25. This morning there was about three inches of new snow on the ground, and the cliffs back of the house are beginning to look white. About 2 P. M. huskies were seen coming across the bay, and a half-hour later they had arrived,—Kayunah, his "koonah" (wife) and three piccaninnies, and Arrotochsuah, his koonah and one piccaninny. Arrotochsuah's koonah was very much amused at me, and kept screaming "Chimo koonah!" (Welcome woman!) until I said " Chimo! Chimo!" and then she laughed and laughed. The other woman was more quiet. These Eskimos are much cleaner and more presentable people than Ikwa and his family. Later in the evening I gave each woman two needles, a cake of soap, and a box of matches. Arrotoch-suah's koonah presented me with a spoon made by herself from a piece of walrus tusk, and used by her piccaninny, Magda, a boy about twelve years old, ever since he could feed himself. In return I gave the boy a looking-glass, and I made a similar present to Kayunah's smallest. Mr. Peary allowed all hands to sleep on the floor in the boys' room. It is amusing to listen to the conversation between our men and the huskies. In one instance the boys could not quite make out whether a man had died from eating walrus or the walrus had eaten him, etc.

Monday, October 26. To-day is the last day the sun will be above the horizon until February 13th.

CHAPTER VII

ESKIMO VISITORS

Our Visitors Leave for their Homes — Departure of a Party to Build a Stone Hut in Tooktoo Valley — Arrival of the Most Northerly Family in the World — The Last Hunting-party of the Season Goes to Five-Glacier Valley — Still the Natives Come — Mama's Birthday — Finishing Touches to our Winter Quarters — Eclipse of the Moon — Beginning of the Winter Routine — Matt Installed as Cook — Thanksgiving.

Wednesday, October 28. Yesterday Nowdingyah and his piccaninny, a little girl about two and a half years old, put in their appearance. The child was nicely dressed in a blue-fox "kapetah" (overcoat) and seal cap trimmed with fox, but she was not as pretty as Kayunah's little one. I gave her a looking-glass, too, which amused her father as much as it did the child. After supper Mr. Peary brought out his reading-glass, and Arrotochsuah's wife immediately said she had seen a white man have one at the northern settlement of Etah, and she showed us how he had used it as a burning-glass. We are all curious to know what party of white men she had seen. The whole evening till midnight was spent in taking flash-light photographs of the Eskimos and ethnological measurements of Kayunah.

Our Eskimo visitors left for their homes this morning. At noon the boys, with Dr. Cook in charge, started for Five-

Glacier Valley to hunt reindeer and to bring the cached venison down to the edge of the ice, where Ikwa will call for it
in a few days and bring it back on the sledge. The boys
will then proceed to the head of the bay, and under Dr. Cook's
direction build a stone igloo for the use of the inland ice party
next spring. About three o'clock Matt returned for a tin of
biscuits which had been forgotten, and informed us that Verhoeff had frozen his nose and face severely, and that Astrup's
cheeks had also been nipped. The temperature was – 10°,
and a fresh southeaster was blowing across the bay. Ikwa and
Mané came in this afternoon and added quite a number of
words to our Eskimo vocabulary; the former also gave us an
account of the murder of his father by tatooed natives while
out after bear off Saunders Island.

Saturday, October 31. Ikwa started this morning with the
sledge and dogs for Arrotochsuah's igloo, where he expects
to get a load of hay. About 2 P. M., while we were out, Mr.
Peary shoveling snow against the wall, we saw a dark object
on the ice, and with the aid of the glass made out a sledge
and two people, but they did not seem to get any nearer, and
in a short time disappeared. About six they arrived — Annowkah, his wife M'gipsu, and an awful-looking baby of about
two months. They came from Nerki, a place beyond Arrotochsuah's, two days' journey from Redcliffe. They are cleaner
and more intelligent-looking than any natives we have yet
seen. In conversation we discovered that they were the most
northerly family of Greenland, and consequently of the globe.

Mr. Peary and I are having great times keeping house by ourselves; he brings in the snow for water, the coal and coal-oil, and keeps watch during the night, while I cook, wash dishes, sweep (without a broom — the only article of importance that was overlooked in the preparations for our Arctic journey), and look after Mané, who is here with her two children working on the reindeer skins. We shall not be sorry when the boys return and take some of these duties off our shoulders.

Thursday, November 5. Jack is the father of eight jet-black pups. The days are only a few hours long now, but the dark-

Prepared for Winter.— My South Window.

ness is not yet the darkness of a winter night at home. Mr. Peary's leg is improving steadily, and he seems more like himself. The strain has told on both of us, and I am glad it is over. He put up his writing-desk yesterday, and our room

is almost fixed for the winter, and looks very cozy. We have
been busy putting up the rest of the blankets in our room,
and have closed the side window and one half of the end
window. As daylight has almost entirely departed this will
make no difference in the amount of our illumination, and the
room will be much warmer, although thus far we have had no
cause to complain, the thermometer not having registered
below 16° at any time.

Our house is by no means a palace, nor do its interior fix-
ings even remotely suggest luxury. We have two rooms, the
smaller of which, measuring twelve feet by seven and a half,
has been reserved for Mr. Peary and myself, while the larger,
of not quite double the size, is used as the general "living-
room," besides affording sleeping-quarters to the boys. A
dining or "mess" table, a few rude chairs, a book-case, and
the "bunks" built to the east wall, constitute the furniture,
of which it can in truth be said there is no superabundance.
The red blanketing which has been tacked all over the inside
walls and the ceiling, seven feet overhead, imparts a warm
feeling to the interior, and relieves what would otherwise be a
cheerless expanse of boards and tar paper. Our stove in the
partition-wall between the two rooms is so placed as to give a
goodly supply of heat to the lowest stratum of the atmosphere.

The shell of the house is made of inch boards, lined inside
and outside with two-ply and three-ply tarred paper, which
is made to fit as nearly air-tight as possible. To the inside of
the ten-inch rafters and posts we have nailed a lining of heavy

cardboard, which forms a support to the blanketing, besides making a complete inner shell of its own. Between the two shells there is free air space, which will greatly help to retain the warmth in the rooms.

A stone wall has been built around the house four feet away from it, and on it we shall store our boxes of provisions, and then stretch a canvas cover over to the roof of the house. Our corridor will thus be sheltered as well as the house, and even in the most inclement weather we shall be able to breathe pure air and have outdoor exercise. With the first heavy snow everything will be plastered over with this natural fleece, and cold though it may be on the outside, we hope to keep quite comfortable within.

Saturday, November 7. To-day has been reception day. We have to-night seventeen huskies in our camp, and I don't know how many dogs; if I were to judge by the howling and yelping, I should say at least fifty. I have been under the weather for the last two days, but feel better to-night.

Sunday, November 8. We generally devote Sunday to sleep; the boys, except the watchman, turn in right after breakfast and sleep till lunch. We have a cold supper, which saves me the trouble of cooking Sunday afternoon. We usually have pemmican and cranberry sauce, salmon, hot biscuits, chocolate, and fruit. Arrotochsuah and his family moved into a snow igloo to-day.

Monday, November 9. Mama's birthday. My thoughts have been at home and with her all day, and I am sure she has thought of me. I do not even know where she is. In my mind I have seen sister Mayde at work on something mysterious for the past week. I must try to put my mind on something else or I shall have a spell of homesickness. I placed a bamboo pole across the front of our bed and draped the two United States flags (one belonging to the National Geographical Society of Washington, and the other to the Philadelphia Academy of Natural Sciences) *à la portière* across the front; then on the wall just beside my place I have hung the photographs of my dear ones.

Saturday, November 14. Very little worthy of note has happened this week. My daily routine is always the same; I take my coffee in bed, then get lunch for my family, take a walk afterward, usually with Mr. Peary, then sew or read, and at four o'clock begin to get dinner. Last Thursday Gibson initiated Frank into dragging a load of ice from the berg to the house. Yesterday was lovely and clear, and the full moon which we have throughout the twenty-four hours, made it as bright as day. Our walk to-day was to the berg, a mile distant (as

Frank.

measured by our newly finished odometer wheel), and return — the first long walk Mr. Peary has taken; his leg did not feel any worse for the trip, but was considerably more swollen at night.   Frank to-day for the first time behaved very well in hauling ice.

Sunday, November 15.   This has been a lovely day.   How much I should like to take a peep at the home folks! To-night we have had the eclipse of the moon.   It was first noticed about 7.30, and Mr. Peary watched it carefully, making observations with his transit and chronometer.   About nine o'clock Arrotochsuah arrived from Netchiolumy,[1] on Barden Bay, accompanied by one of his sons and another young man. The first we immediately nicknamed the "Smiler," and the other the "Villain," owing to the expressions on their faces.

Tuesday, November 17.   Yesterday was an exceptionally fine day, beautifully moonlit.   The "Villain" of Netchiolumy has a sledge made of the boards which Dr. Cook traded for a tupic when the "Kite" stopped at the settlement in July. This morning Ikwa introduced a rather clean-looking native from Omanooy, a place this side of Akpani, on Saunders Island; his name is Kioppadu. Our sewing progresses slowly, Arrotochsuah's wife, whom we had installed as seamstress, being too old to prepare the skins by the time-honored native method of chewing.   Matt got supper to-night, and will from now until May 1 prepare all the meals under my supervision.   This gives me more time to myself, besides not confining me to the house.   It was no easy task for me to cook for six boys, and for such appetites.

1 Erroneously called by most geographers Ittiblu.

Thursday, November 19. We have had our first real winter snow-storm to-day. The wind whistled, and the snow was driven into every crack and crevice. Just before noon Kayunah and family came; Makzangwa, his wife, is going to chew skins for us. They will live in the snow igloo, having brought all their household effects with them; these consist of the soapstone blubber lamp or stove, a reindeer skin as a coverlet for the bed (which is merely a bundle of hay on some pieces of board given them by us), a few rabbit and gull skins for wraps for the feet, and a sealskin to put against the wall behind the bed. When these articles are put inside the igloo, their house is furnished.

Saturday, November 21. A clear day; the stars are twinkling and the air is delightful, but one must exercise to keep warm. Since Matt does the cooking, I take long walks every day, and find them very agreeable. We had a general house-cleaning to-day, and will have it now every Saturday. We have been obliged to dismiss the Eskimos from the living-room during meal-time, as their odor is too offensive.

Sunday, November 22. Kayunah came in this morning, and said that our coffee and biscuit made his family sick, and as they had no more seal meat they must go home. Mr. Peary gave them permission to help themselves to the walrus stacked up behind our house, and the Eskimo was satisfied. Ikwa and Kyo (Kioppadu) have gone over to the settlement of Igloodahominy, on Robertson Bay, after blue foxes.

6

Monday, November 23. It grows gradually darker every day. To-day at noon it was impossible to read ordinary print by daylight. Mr. Verhoeff went on the cliffs to look at his thermometer, and found that it read higher than those at Redcliffe. Ikwa and his brother returned about noon without foxes or game of any kind. We had a faint aurora this evening. On the whole I am very much disappointed in the auroras; I thought we should have very beautiful displays in the Arctic regions, but it seems that we are too far north of the magnetic pole.

Wednesday, November 25. The days are rather unsatisfactory, although I keep busy all day sewing, mending, rearranging my room, etc. When I sum up at bedtime what I have accomplished, it is very little. Mr. Peary and the boys are busily at work on some test sledges. This afternoon Annowkah and M'gipsu returned, bringing with them a twelve-year-old girl, named Tookymingwah, whose father was dragged under the ice and drowned a few weeks ago by an infuriated "oogzook" seal *(Phoca barbata?)* which he had harpooned. She has a mother and two sisters, who will be here soon.

Mr. Peary issued the Thanksgiving proclamation, and I have been busy getting things ready for the Thanksgiving dinner, which I told Matt I would prepare. Our cooking and baking is all done on oil-stoves; since I have only three ovens I baked my pies to-day, as I shall need all the stoves and ovens to-morrow. This forenoon I went out to our berg, accompanied by Mr. Peary and my two Newfoundland dogs, after a load of ice. It is rather a novel idea to me, chopping ice from the stately icebergs and melting it for drinking and cooking purposes.

Thursday, November 26. Thanksgiving day, and all work is suspended. Before lunch I went down to Cape Cleveland with Mr. Peary to see how much daylight still remains toward the south. The sky was tinged with rose near the southern horizon, and the moon was just coming up from behind Northumberland Island. How strange it is that while we have no sunlight whatever, we know that at home they are having day and night just as usual! The temperature was 12½° F. Dinner was served at 7 P. M. All the boys wore American clothing, and the room was draped with the Stars and Stripes.

CHAPTER VIII

ARCTIC FESTIVITIES

Creeping Toward the Winter Solstice — Household Economy — The Holidays —
Christmas Amusements — Christmas Dinner to the Natives — New-Year Festivities — Moonlight Snow-shoe Tramps — Reception in the South Parlor.

Wednesday, December 2. Thanksgiving has come and
gone. We had a very pleasant time, and enjoyed our dinner as much as any one at home. The only difference between day and night at Redcliffe is that during the day in
addition to the bracket-lamps we have a large Rochester
lamp burning. The huskies, as we continue to call the natives, have named it the "mickaniny sukinuk" (baby sun).
Matt lights it at 8 A. M., and the officer on watch puts it out
at 10 P. M. Mr. Peary has made a rule that no member of
the party, unless ill, shall occupy his bunk between the hours
of 8 A. M. and 7 P. M. He has also changed from the four-
hour watches to twelve-hour watches; thus one man has the
night watch for a whole week, and during this time sleeps in
the daytime, and one man has the day watch. At the end
of a week these two men are relieved by two others. The
boys think they like this arrangement very much better.
The native whom Ikwa brought back with him from Keati

OUR FRIENDS ABOUT REDCLIFFE.

is named Mahoatchia, and Ikwa says that he and the one-eyed bear-hunter, Mekhtoshay, of Netchiolumy, exchange wives with each other every year. It is interesting to note that these two men are the only ones in the tribe who indulge in this practice, yet the other men seem to think it all right; but the women are not at all satisfied with this social arrangement.

If some of our dear ones at home could look down upon us now they would be surprised to find how comfortable and contented we are. Everybody is busily engaged in getting the equipment and clothing ready for the long spring sledge journey over the inland ice. Mr. Peary gives me an idea of what kind of garments he wants, and I am making experimental outfits out of canton flannel, which, when satisfactory, will be used as patterns by which the skins will be cut, thus avoiding the chance of wasting any of the valuable furs. While I am at work on this, two native women, M'gipsu, wife of Annowkah, with her baby on her back, and Tookyming-wah, the twelve-year-old girl, are both sitting tailor-fashion on the floor, chewing deerskins. The native method of treating the skins of all animals intended for clothing, is first to rid them of as much of the fat as can be got off by scraping with a knife; then they are stretched as tight as possible, and allowed to become perfectly dry. After this they are taken by the women and chewed and sucked all over in order to get as much of the grease out as possible; then they are again dried and scraped with a dull implement so as to break the fibers

6*

making the skins pliable. Chewing the skins is very hard
on the women, and all of it is done by them; they cannot
chew more than two deerskins per day, and are obliged to
rest their jaws every other day.

Kyo, Ikwa's brother, and Annowkah come in occasionally
and scrape some of the skins after they have been chewed.
Kyo especially tries to make himself useful. He presents
rather a comical appearance in his bearskin nanookies and blue
guernsey given him by one of the boys. Every time he sees
any shavings or other trash on the floor he seizes the broom,
made by him out of the wings of eider-ducks, and sweeps
it up. Mr. Peary and the boys are carpentering from morning till night, and every day we assure one another that we do
not mind the Arctic night at all; but I don't think that any of
us will object to seeing the sun again.

Thursday, December 10. A whole week has passed since
I wrote in my journal. We have had one or two very disagreeable days, the wind making it too unpleasant for my
daily walk.

We have been busy working on the fur outfits. I have succeeded in getting satisfactory patterns for Mr. Peary; Mané
and M'gipsu are sewing. The former is a poor sewer, but
M'gipsu is very neat as well as rapid, and I have suggested to
Mr. Peary that he offer her an inducement if she will stay and
sew until all the garments are completed. She understands
us and we understand her better than any of the other natives,
including Ikwa and Mané, although they have been with us

fully ten weeks longer. I hope it is not a case of new broom, and that she will wear well. The little girl Tookymingwah, whom we all call "Tooky," is a neat little seamstress, but is not very rapid. A few days ago her mother, named Klayuh, but always called by us the "Widow," arrived with her two younger daughters, the youngest about five years old. I asked her if she had only the three children, and she burst into tears and left the house without answering me. Turning to M'gipsu, I asked her what it meant, and she said it was "peuk nahmee" (not well) for me to

M'gipsu Sewing.

ask Klayuh about other children. When I insisted upon knowing why, she took me aside and whispered that Klayuh had just killed her youngest child, about two years of age, by strangling it. She went on to explain that it was perfectly right for Klayuh to do this, as the father of the child had been killed, and she could not support the children herself, and no man would take her as a wife so long as she had a child small enough to be carried in the hood. I asked her if this was always done, and she said: "Oh, yes, the women are compelled to do it."

Mr. Peary has spoken to M'gipsu about staying at Redcliffe as seamstress, and she is delighted at the opportunity. When Ikwa heard of this arrangement he rushed in and wanted to know why he was "no good" for Peary, and why Mané could not do the sewing, and said that if Peary preferred Annowkah and M'gipsu he would pull down his igloo and take his family back to Keati. It was some little time before we could quiet him and make him understand that we needed more than one woman to sew all of the clothing.

The last three days have been particularly busy ones for me, as Matt has been sick in bed with something like the grippe, and I have had the cooking to do in addition to the sewing. The poor fellow has had an uncomfortable time, but the doctor says he will be all right in a day or two.

Our house looks like a huge snow-drift from a little distance, so completely is it covered with snow. The whole village presents the appearance of a series of snow-mounds of various sizes. We have five snow igloos inhabited by the natives, besides a storehouse, an experimental snow-house, and some dog-houses, all built of blocks of snow. Just at present we are getting quite a little amusement out of two young natives from Cape York, who express the same surprise at us and our mode of living as the country boy does the first time he comes to a city. They are dressed in new suits throughout,—kamiks, bearskin nanookies, foxskin kapetahs, and birdskin shirts,— and so the boys have nicknamed them the "Cape York dudes." The younger one, Keshu, is a stepbrother of Klayuh, and he has brought her the sad tidings that their father is very sick and will probably never get well again. I should not be surprised if she would return to Cape York with them.

Monday, December 21. The dark night is just half over; to-day is the shortest day. So far the time has not seemed very long, but I am afraid before we have had many more dark days we shall all think it long enough. I have done nothing as yet toward celebrating Christmas, but I want to make some little thing for Mr. Peary. As far as the boys are concerned, I think an exceptionally good dinner will please them more than anything else I could give them. M'gipsu has made a pair of deerskin trousers for one of the boys, and has also completed a deerskin coat. She is now at work on a deerskin sleeping-bag, which is to be fastened about the neck of the occupant, over a fur hood with a shoulder cape, which I am endeavoring to fashion.

She is sitting on the floor in my room (an unusual honor), and her husband, Annowkah, comes in as often as he can find an excuse for doing so. He frequently rubs his face against hers, and they sniffle at each other; this takes the place of kissing. I should think they could smell each other without doing this, but they are probably so accustomed to the (to me) terrible odor that they fail to notice it.

I dislike very much to have the natives in my room, on account of their dirty condition, and especially as they are alive with parasites, of which I am in deadly fear, much to the amusement of our party. But it is impossible for the women to sew in the other room, where the boys are at work on their sledges and ski, so I allow two at a time to come into my room, taking good care that they do not get near the bed. At the end of their day's work, I take my little broom, which is an ordinary whisk lashed to a hoe-handle, and sweep the room carefully. The boys have made brooms out of the wings of ducks and gulls, which are very satisfactory, there being only the bare floor to sweep; but I have a carpet on my floor, and the feather brooms make no impression on it, so I am compelled to use my little whisk. It answers the purpose admirably, but it takes me twice as long as it would otherwise have done. After the room has been thoroughly swept, I sprinkle it with a solution of corrosive sublimate, given to me by the doctor, and in this way manage to keep entirely free from the pests. Both Mr. Peary and myself rub down with alcohol every night before retiring as a further protection against these horrible "koomakshuey," and we are amply repaid for our trouble. Matt has entirely recovered from his sick spell, and has again taken charge of the cooking.

I was right in my surmise about the widow; she accompanied the "dudes" to Cape York, taking her three children with

her. Kyo also left at the same time for his home at Omanooy.
He says he will return in ten days with a load of deerskins
which he has at his igloo. Mr. Peary loaned him two of his
dogs, and has promised him ammunition in exchange for
the deerskins. We are anxious to see what kind of a gun he
has; he says he got it from an old man who had received it
from a white man long ago.

We have had a great house-cleaning in honor of the approaching holidays. I have replaced the cretonne curtains at
the bottom of my bed, wash-stand, bookcase, and trunk, with
new ones, and have put fresh muslin curtains at my windows.
The boys have cleaned the large room, taking all superfluous
lumber and tools out, and have even scrubbed the floor. The
natives think we are crazy to waste so much water. Poor
things, they think water was made only for drinking purposes.

Saturday, December 26. Just after I made the last entry
in my journal, one of the boys reported that the tide-
gage wire was broken. Mr. Peary, Verhoeff, and Gibson
went out to put it in commission. After about an hour Verhoeff rushed into the house calling, " Doctor, Doctor, come
out to the tide-gage as quick as you can!" The doctor,
whose turn it was to be night-watchman, and who was therefore asleep at this hour, tumbled out of his bunk and into his
clothes, and made a rush for the tide-gage. I was lying in
my bed suffering from the effects of a sick-headache; but never
having fully recovered from the shock caused by Mr. Peary's
accident in Melville Bay, and realizing that he was not yet

quite sure of his injured limb, the thought flashed across my mind that something had happened to him. No sooner did this idea occur to me than it became a settled fact, and in less time than it takes to tell I had thrown on my wrapper and kamiks, caught up a steamer-rug to throw about me, and was on my way down to the tide-gage. As I ran down the beaten path, I could see the light of the little bull's-eye lantern flashing to and fro in the distance. It was as dark as any starlight night at home, although it was early in the evening, and not any darker now than it had been at noon. I could hear the low buzz of conversation without being able to distinguish any voices, and the figures seemed all huddled together. My whole attention was absorbed by this little group, and I did not properly watch my path; consequently I stumbled, then slipped and lost my footing, falling astride a sharp ridge of ice on the ice-foot. For an instant I could not tell where I was hurt the most, and then I discovered that I could move neither limb, the muscles refusing to do my bidding. I next tried to call Mr. Peary, whose voice I could now distinctly hear, but I could utter no sound. Then I lost consciousness. The next thing I knew, I was lying on the same spot in the same position. The little group, not more than sixty yards away, were laughing and talking; but I was unable to raise my voice above a hoarse whisper, and could in no way attract their attention, so interested were they in their work of raising the tide-gage anchor. I was clothed in such a way that lying out on the ice with the temperature eighteen degrees below zero was anything but comfortable. I found that by great exertion I could move myself, and by doing this a little at a time, I gradually got on my hands and knees and crawled back to the house. As the whole distance was up-hill and every movement painful, I was obliged to make frequent stops to rest. At last I reached my room and had just strength enough left to drag myself upon the bed. I noticed by the clock that I had been absent thirty-five minutes. On examination it was found that I was cut and bruised all over, but the doctor declared that I was not seriously hurt; but even now I have not entirely recovered from the effects of the fall.

The day before yesterday was spent in decorating the interior of our Arctic home for the Christmas and New-Year festivities. In the large room the ceiling was draped with red mosquito-netting furnished by Mr. Gibson. Dr. Cook and Astrup devised wire candelabra and wire candle-holders, which were placed in all the corners and along the walls. Two large silk United States flags were crossed at one end of the room, and a silk sledge-flag given to Mr. Peary by a friend in Washington was put up on the opposite wall. I gave the boys new cretonne for curtains for their bunks. In my room I replaced the portières, made of silk flags, with which the boys had decorated their room, by portières made of canopy lace, and decorated the photographs of our dear ones at home, which were grouped on the wall beside the bed, with red, white, and blue ribbons. This occupied us all the greater part of the day. About nine o'clock in the evening

Mr. Peary made a goodly supply of milk-punch, which was placed upon the table, together with cakes, cookies, candies, nuts, and raisins. He gave each of the boys a book as a Christmas gift. We spent the evening in playing games and chatting, and at midnight Mr. Peary and I retired to our room to open some letters, boxes, and parcels given to us by kind friends, and marked, " To be opened Christmas eve at midnight." I think our feeling of pleasure at the many and thoughtful remembrances was clouded by the feeling of intense homesickness which involuntarily came with it. It was the first Christmas in my life spent away from home, and for the first time since the little " Kite " steamed out of Brooklyn I felt how very far away we are from those we love and who love us. I shall never forget the thoughtful kindness of Mrs. Beyer, wife of the governor of Upernavik, to a perfect stranger. Although she is obliged to get all her supplies from Denmark, and then order them a year in advance, out of her slender stock she had filled a large box with conserves, preserves, bonbons, spice-cakes, tissue-paper knickknacks for decorating the table, and very pretty cards wishing us a merry Christmas. Mr. Peary had carved for me two beautiful hairpins, and I made a guidon out of a silk handkerchief and a piece of one of my dresses, to be carried by him on his long journey over the ice-cap to the northern terminus of Greenland.

Yesterday — Christmas morning — we had a late breakfast, and it was very near noon before all the inmates of Redcliffe were astir. I had decided to have an early dinner, and then to

invite all our faithful natives to a dinner cooked by us and
served at our table with our dishes. I thought it would be as
much fun for us to see them eat with knife, fork, and spoon
as it would be for them to do it.

While I was preparing the dinner, most of the boys went
out for a walk, "to get a good appetite," they said. After the
table was set, Astrup placed a very pretty and cleverly de-
signed menu-card at each plate. Each card was especially
appropriate to the one for whom it was intended.

At 4.30 P. M. we all sat down to our "Merry Christmas."
The dinner consisted of

Salmon *à la* can.
Rabbit-pie with green peas.
Venison with cranberry sauce.
Corn and tomatoes.
Plum-pudding with brandy sauce.
Apricot-pie.
Pears.
Candy, nuts, raisins.
Coffee.

We arose from the table at half-past seven, all voting this to
have been the jolliest Christmas dinner ever eaten in the
Arctic regions. After Matt had cleared everything away, the
table was set again, and the Eskimos were called in. Ikwa
and his family sent regrets, as they had just returned from a
visit to Keati, and were too tired to put on "full dress" for a
dinner-party. We therefore had only two of our seamstresses,

M'gipsu and Inaloo, with us; in place of Ikwa and his wife we invited two visitors, Kudlah and Myah. We had nicknames for all the natives. Ahngodegipsah we called the "Villain" on account of the similarity of his expression, when he laughed,

Christmas Dinner to the Natives.

to that of the villain on the stage. His wife, Inaloo, talked so incessantly that she at once received from the boys the nickname of the "Tiresome." M'gipsu was called the "Daisy" because she could do anything she was asked to do. Her husband, Annowkah, we knew as the "Young Husband"; Kudlah was called "Misfortune"; and Myah was known as the "White Man." The "Villain" was put at the head of the table

and told that he must serve the company just as he had seen
Mr. Peary serve us. The "Daisy" took my place at the foot of
the table, her duty being to pour the tea. The "Young Husband" and "Misfortune" sat on one side, while "Tiresome"
and the "White Man" sat opposite. Their bill of fare was as
follows:

Milk-punch.
Venison-stew, corn-bread.
Biscuit, coffee.
Candy, raisins.

It was amusing to see the queer-looking creatures, dressed
entirely in the skins of animals, seated at the table and
trying to act like civilized people. Both the "Villain" and
the "Daisy" did their parts well. One incident was especially
funny. Myah, seeing a nice-looking piece of meat in the stew,
reached across the table, and with his fork endeavored to pick
it out of the dish. He was immediately reproved by the
"Villain," who made him pass his mess-pan to him and then
helped him to what he thought he ought to have, reserving,
however, the choice piece for himself. They chattered and
laughed, and seemed to enjoy themselves very much. Both
women had their babies in their hoods on their backs, but this
did not hinder them in the least. Although at times the
noise was great, the little ones slept through it all.

M'gipsu watched the cups of the others, and as soon as she
spied an empty one she would say: "Etudoo cafee? Nahme?
Cafee peeuk." (More coffee? No? The coffee is good.) Fin-

7

ally at ten o'clock the big lamp was put out, and we told them it was time to go to sleep, and that they must go home, which they reluctantly did.

To-day has been a rather lazy day for us all, and now at 11 P. M. Mr. Peary, Dr. Cook, and Matt have just come in from a visit to the fox-traps about two miles distant. On the return they indulged in a foot-race, and when they came in they looked as if they had been dipped in water. The perspiration ran in streamlets down their faces. This trip has encouraged Mr. Peary very much in the belief that by next spring his leg will be just as good as it ever was.

Saturday, January 2, 1892. I have been lazy about writing up my notes lately, but now I shall turn over a new leaf. 1891 has gone; what will 1892 bring? I don't think I want to know. Better take it as it comes, and hope for the best. The "Villain" and his wife have gone to their home in Netchiolumy, Myah and Kudlah also have left us, and, with the exception of Keshu (alias the "Smiler") and his wife, all of our Eskimo visitors have departed; Ikwa and family and Annowkah and family remain, but they are not considered company at Redcliffe.

The sun is surely coming back to us, for at noon now we have a perceptible twilight, and the cliffs opposite Redcliffe can be plainly seen. Since December 29 the weather has been very disagreeable, and we have considerable new snow. The whole week has been a semi-holiday. Almost every day I have been out for a snow-shoe tramp, and I have rather enjoyed it in spite of the wind, which is just high enough to be disagreeable.

On the 30th I issued cards of invitation for an "At home in the south parlor of Redcliffe, December 31, from 10 P. M. 1891 to 1892." The day was a thoroughly Arctic one, and I was glad that my guests would not have far to come. All day I was busy preparing for company. I had to manufacture my own ice-cream without a freezer, bake my own cake and crullers, and set everything out on an improvised sideboard. At 9 P. M. I dressed myself in a black silk tea-gown with canary silk front, covered and trimmed with black lace, cut square in the neck and filled in with lace, and having lace sleeves. At ten my guests began to arrive. The invitations were limited to the members of the North Greenland Expedition of '91 and '92, and they all looked especially nice and very much civilized, most of them actually sending in their cards. They were all dressed in "store clothes," although one or two clung to their kamiks. I had no chairs, so each guest was requested to bring his own. Mr. Peary sat on the bed, while I occupied the trunk. I spent a very delightful evening, and I think the boys enjoyed the chocolate ice-cream and cake. At midnight we all drank "A Happy New Year" in our Redcliffe cocktail, and then my guests departed. All this time the wind was howling and moaning, and the snow was flying, while the night was black as ink, not a star being visible. More than once during the evening, when a particularly heavy gust swept down from the cliffs and fell against our little house with a shriek, the contrast between inside and outside was forced upon us.

The next day we had a late breakfast, and then two of the boys went out to lay off a course for the athletic games which they had been discussing for some time. The weather was so bad that I did not go out to witness them, but let Matt go, and prepared our New-Year's dinner alone. This time Mr. Peary decided that he would give the natives the materials for their own New-Year's dinner and let them prepare it themselves. They were given eider-ducks, reindeer legs, coffee, and biscuit. We have quite a batch of new Eskimos, among them two men from Cape York, who are almost as tall as Mr. Peary, and whom we call the "giants." They have quite a number of narwhal tusks to trade, and are determined to have a rifle for them, but I hardly think they will get it.

CHAPTER IX

THE NEW YEAR

The New Year Ushered in with a Fierce Storm — Return of the Noon Twilight — We fail to feel the Intense Cold — Native Seamstresses and their Babies — Some Drawbacks to Arctic Housekeeping — Peculiar Customs of the Natives — Close of the Winter Night.

Saturday, January 9. The storm which began December 29 has continued until this morning. Now it looks as though it might clear off. The new snow is about twenty-four inches deep on a level, and there are drifts as high as I am.

Fortunately we had a good ice supply on hand, and no native visitors, for they drink twice as much water as we use for cooking, drinking, and toilet purposes combined. The boys have been busy on their individual ski and sledges; Mr. Peary has been fitting and cutting fur clothing and sleeping-bags; and the "Daisy" has been sewing as hard as she can. The wind is still blowing in squalls, and of course the snow is still drifting, but the moon came out for a little while to-day, and we think and hope the storm is over.

Monday, January 11. At last clear and cold, and the twilight is very pronounced in the middle of the day. Everybody is still busy sewing or carpentering. Each one of the party is desirous of having his ski lighter and stronger than those of the others, except Verhoeff, whose whole interest is divided between the thermometer and the tide-gage. The words of the physicians on board the " Kite " six months ago have come true — Mr. Peary's leg is practically as sound as it ever was.

Saturday, January 16.   During the last week we have had beautiful weather—calm, clear, and cold. Every day we have a more decided light, and I take advantage of it by indulging in long snow-shoe tramps. I can walk for hours without tiring if a single snow-shoer has gone before me; but if I attempt to

break the path alone I soon get exhausted. I have been busy making foot-wraps out of blanketing, and have also made myself some articles of clothing out of the same material. We find that mittens made out of blanketing and worn inside the fur mittens absorb the moisture and add to the warmth and comfortable feeling.

My room has looked more like a gun-shop than anything else for the

In my Kooletah.

last few days; Mr. Peary has been putting a new spring in his shot-gun and overhauling an old rifle.

Sunday, January 17.   To-day at 2 P. M. Mr. Peary and I went out for our tramp. The temperature was −45°, and the only chance to walk was along the pathway made through the twenty-inch depth of snow three quarters of the way to the iceberg. It is astonishing how little I feel these low temperatures: Mr. Peary, however, always sees that I am properly protected. In many of the little details I should be negligent, and would probably suffer in consequence, but I have to undergo an inspection before he will let me go out.

The daylight was bright enough to-day to enable us to read ordinary print, and we feel that ere long we shall have the sun with us again for at least a portion of the twenty-four hours. We stayed out only half an hour, but my dress for about two feet from the bottom was frozen stiff as a board, my kamiks were frozen to the stockings, and the stockings to the Arctic socks next my feet; yet I have felt much colder at home when the temperature was only a little below the freezing-point.

The remainder of the day we spent in marking, clipping, and sorting newspaper cuttings. This occupation we found so interesting that we prolonged it until after midnight.

Monday, January 18. The day has been bright and calm. Mr. Peary, with Dr. Cook and Astrup, took his first snow-shoe tramp of the season, and went nearly to the berg. This is the first time the broken leg has been given such vigorous exercise, but it stood the strain remarkably well. I have been busy on the sleeping-bag cover all day. I find it very inconvenient, not to say disagreeable, sewing in a temperature of 44°; but as I am dependent on the stoves in the other room for my heat, it cannot be helped. Verhoeff has a mania for

saving coal, and keeps everybody half frozen. He kept the
fire to-day on six tomato-cans of coal. Water spilled near the
stove froze almost instantly.

Tuesday, January 19. Somewhat cloudy to-day, but after
lunch Mr. Peary and I went out to the berg on snow-shoes. I
did not get a single tumble, and Mr. Peary said I managed my
snow-shoes very well. I was as warm as any one could wish to
be, although the thermometer registered 44° below zero. We
took our time, not hurrying at all, and so prevented perspiration, which always makes one uncomfortable in these low
temperatures. I had no shoes or kamiks on, only the deerskin stockings, and a pair of long knit woolen ones over
them, yet my feet were warmer than ever before on these
outdoor tramps.

Thursday, January 21. A clear and perceptibly lighter
day than yesterday; indeed, it seems as if it grew lighter now,
a month after the shortest day, much more rapidly than it grew
darker a month before the shortest day. Mr. Peary, the doctor, and Astrup started a path with their snow-shoes toward
Cape Cleveland, and made about half the distance. The doctor and Astrup took our sledge, the " Sweetheart," to the iceberg, intending to bring in a load of ice, but as they reached
the berg they heard the howling of dogs ahead of them and
saw a dark object on the snow some distance away. They
started for it, and found a party of huskies plowing their way
through the snow. The party consisted of Keshu, his wife
and child of three years, his brother, Ahninghahna, older than

AMPHITHEATRE BERG—McCORMICK BAY.

he, and Magda, a boy of twelve. They were on their way
to Redcliffe. They had been staying with Keshu's father,
Arrotochsuah, but as the food was giving out over there, and
as the old people were not able to travel, they thought it
desirable to look elsewhere. They all have frost-bites except
the little child, and were very grateful for the assistance given
them by the doctor and Astrup in getting to the house. They
tell us that they have been on the way for five days and
nights, the distance being about fifteen miles. To-night the
woman was photographed, and her portrait added to our ethnological series.

Friday, January 22. Another clear, cold day; the temperature, —39°. The addition of the new Eskimos makes the
settlement much more lively. In the house I wear a knit
kidney-protector, a Jaros combination suit, two knit skirts, a
flannel wrapper, and a pair of knit stockings, together with a
pair of deerskin ones in place of kamiks. When going out I
only add my snow-shoes, my kooletah (great fur overall), and
muff. In this rig I can stay out and walk for hours, and feel
more comfortable than I have felt while shopping in Philadelphia or New York on a winter's day. This evening Mané
No. 2 (wife of Keshu) and M'gipsu have been at work in my
room, both sitting flat on the floor, the former cutting and fitting two pairs of kamiks for us from a skin brought here by
herself, for which she will receive a clasp-knife. The bargain
pleases her greatly. These women are both good sewers, and
it would interest some of our ladies to watch them at their

work. They, as well as all the other native women, usually
take off their kamiks and stockings while in the house, so that
almost the entire leg is bare, their trousers being mere trunks.
They sit flat on the floor, using their feet and legs to hold
the work, and their mouths to make it pliable; the thimble is
worn on the forefinger, and they sew from right to left. The
thread is made as they need it by splitting the deer or narwhal
sinews and moistening them in the mouth. While at this
work the babies are being continually rocked or shifted on
their backs without the aid of the hands. The children are
carried in the hood constantly, whether awake or asleep, for
the first year, and only taken out when fed. They are tiny,
ugly creatures, and until they are able to walk never wear
anything but a sealskin cap which fits close about the face,
where it is edged with fox, and a foxskin jacket reaching to
the waist.

Saturday, January 23. I cleaned "house," which means
our little room, seven by twelve. This in itself would be no
task, but we have no brooms, and every inch of my floor is
swept with a whisk-broom and on my knees. As I have only
one whisk, and that a silver-handled one, I can afford to sweep
thoroughly only once a week. I have put an old blanket
down which covers the carpet in the middle of the room,
where all the walking and working is done. This blanket is
shaken every day and the room brushed up, giving us a fairly
clean apartment. I also finished the sleeping-bag cover. Now
at midnight the temperature is $-30\frac{1}{2}°$, and the doctor and

Astrup have taken their sleeping-bags out under the boat as an experiment in sleeping in the open air.

Monday, January 25. A clear, calm day, with the very bright daylight tipping all the bergs and crests of the cliffs with silver. The temperature is −29°, and the landscape is a cold-looking one, but its aspect does not chill us. It is certainly novel to feel so decidedly hot in a temperature of −30°, while my handkerchief freezes stiff be-

A Winter Recreation.— My Cross-matched Team.

fore I get through using it. I have been busy cutting and sewing a flannel lining for my reindeer knickerbockers, for which I utilized my old gray eiderdown wrapper. I also made out a schedule or bill of fare for the week, arranging the *menu* for each day, so as to get the greatest benefit from the patent-fuel stove and save as much oil as possible.

Thursday, January 28.   About five o'clock I was called out to see the brightest aurora we had yet seen.   It extended over us almost due east and west.[1]   This night we succeeded in obtaining an observation of Arcturus.

Friday, January 29.   To-day we went out to the "amphi-theater berg," breaking a new path part of the distance — warm as well as hard work.   This evening, for the first time in our house, one of the women (Mané) stripped herself to the waist; there she sat sewing away, in the midst of a crowd of huskies as well as our boys, just as unconcerned as if she were clad in the finest raiment.   The men do this frequently when it gets too warm for them, but I never saw a woman do it before.   It is true they are nearly always entirely nude in their igloos, and visiting Eskimos, as soon as they enter an igloo, take off every stitch, just as we lay aside our wraps and overcoats at home.   This is done by both sexes.

Sunday, January 31.   Another month has slipped away, and I can say, "One month nearer home."   I must admit I am very homesick at times.   Hardly a night passes that I do not dream of some of my home folks.   The bill of fare which I made out for last week, giving the times for cooking each dish on the patent-fuel stove, worked very well, and I can save about one quart of oil a day; this will be of considerable help to us in case we shall be obliged to go to south Greenland in our boats.   I walked down to the two first fox-traps, but

[1] This was the only aurora observed by us during our entire stay in the Arctic regions which was bright enough to cast a shadow.

found them completely snowed under. In places the snow-crust is hard enough to bear the weight of the body, but oftener one sinks in six or eight inches, and in places the surface snow has drifted considerably deeper. The temperature is about —20°, and it has been thick and dark all day. Yesterday Verhoeff went upon the cliffs and found the minimum thermometer registering only —24° as the lowest for the month, while at Redcliffe we have had it down to —53°. Strange that on the hill-tops it should be so much warmer than here below.

Tuesday, February 2. A beautiful, clear, cold day; temperature, —35°. We now have daylight from ten A. M. until three P. M., while there is a decided twilight from nine to ten and from three to four. We were inspected in daylight by the doctor, and we all show the effects of the long dark night; Mr. Peary and Astrup, being the two fairest ones in the party, look the most sallow. We walked out to the amphitheater berg without snow-shoes. The left-hand column at the entrance to the theater is a massive pillar of ice, like the whitest marble, about a hundred feet high; inside the berg the snow was very deep. The right-hand side of the entrance had recently broken, and tons of the splintered ice were lying around. We saw the new moon one quarter full for the first time over the cliffs to the north, while the glow from the setting sun to the southwest made a most beautiful picture; the tops of the bergs in the distance were completely hidden in the low line of mist rising from the cracks in the ice, which gave them the appearance of long flat rocks in the midst of the snow-plain.

Friday, February 5. This morning all our Eskimo visitors left us, and things are once more running in the old groove. I have not been out for several days in consequence of a sore toe. I have finished blanket sleeves for all the sleeping-bags, and yesterday boiled my first pudding. To-night about eight o'clock noises were heard out on the ice, and in a little while Arrotochsuah and his wife arrived, with one large dog and one puppy. They were very much fatigued, having been five days and four nights on their way over. These old people seem very fond of each other, and share whatever they get. Their food-supply having given out, they are on their way to their son's igloo at Netchiolumy, forty-five miles distant, whither they intend to travel on foot, part of the way through snow two feet deep. The woman, seemingly sixty years of age, says they tumble into the snow every few steps, but up they get and stagger on, and in this way they make the trip with packs on their backs.

Thursday, February 11. Just seven months ago to-day Mr. Peary broke his leg, and he celebrated the event by taking a ten-mile tramp on the bay ice. His leg did not trouble him at all, and did not swell very much. To-day we have been married three years and a half. It seems as if I had been away from home as long as that, and yet it was only eight months on the 6th of February since I left Washington.

Saturday, February 13. We are making preparations to

witness the return of the sun.   Gibson and Verhoeff have erected a snow-house on the ice-cap, and Mr. Peary has invited us all to accompany him to-morrow to the summit, and welcome the reappearing luminary.   My head has been aching very badly all day, and I do not feel in condition to spend the night in a snow-hut, so I shall stay at home and keep house. It will be pleasant to exchange the strange daylights we have been having for weeks—daylights without a sun—for the vivifying glow of direct sunlight.

CHAPTER X

SUNSHINE AND STORM

Return of the Sun — Furious Storm and Inundation at Redcliffe — Repairing the Damage — Verhoeff's Birthday — Fears for Dr. Cook and Astrup — Rescue of Jack — Battling with an Arctic Hurricane — Down with the Grippe — Dazzling March Scenery — The Commander has the Grippe — Astrup and Gibson reconnoiter after Dogs — The Widow returns a Bride — The Snow begins to Melt — Sunning Babies on the Roof.

Sunday, February 14. At home this is St. Valentine's day. Here it is simply Sunday, and for me a lonely one. This morning Mr. Peary, Astrup, and Dr. Cook started for the mountain-top with their sleeping-gear and provisions for two days. The day has been misty, cloudy, and rough. At six A. M. the temperature was 11½°, and at eight it was 33°, with the wind blowing a gale that shook the doors and windows of our little home for the first time since it was really finished. At eight in the evening the mercury had fallen one degree, and the wind was blowing in gusts, but with greater force than before. I am worried about our travelers. Gibson just brought in a piece of ice perfectly wet and covered with wet snow, which shows the effect of the high temperature. He says he can hardly stand up against the wind, but that it is warm, almost balmy. Jack came to the door and whined piteously to be let in, something I have never known him to do before. Now at 10.45 it is raining hard.

Monday, February 15. What a wretched twenty-four hours the past have been! All night the wind blew in violent gusts, sometimes accompanied by wet snow and sometimes by rain. This morning the whole place appears in a dilapidated condition. A thaw has set in, and the water is running in every direction. The inmates of the snow-igloo were forced to leave it, and to-night one could read through its walls, the action of the wind, water, and temperature has worn them so thin. Part of our snow-wall has fallen, or rather melted down, and the water is pouring down the sides of the house into the canvas-covered passages, soaking everything. The thermometer reads 38°, and the wind still blows, while it continues to rain and snow. With Matt's assistance I have moved everything out of the lean-to back of the house, and have had all the cutlery brought in, some of which was already covered with rust. At two o'clock the water began to come in under my back door, and then Gibson, who has the night-watch, and therefore the right to sleep during the day, got up, and with Matt went on the roof and shoveled the snow off to prevent the water from leaking into the house. It was all they could do to keep from being blown down, and in ten minutes both were drenched to the skin. If our little party on the ice have this wind and rain, I do not see what they can do. Their snow-hut will melt over them, and they will be wet and cold, while in such a wind it will be impossible to venture down the cliffs. To-

8

night the temperature has fallen to 33°, but otherwise things are unchanged. At two P. M. the maximum thermometer registered 41½°. This temperature will hardly be equaled at this time in New England.

Tuesday, February 16. A glorious day follows thirty-six hours of violent storm. The sun shines on Cape Robertson and on the snow-covered cliffs east of Redcliffe House. I walked down to Cape Cleveland with Jack, my faithful attendant. The sun had just gone behind the black cliffs of Her-

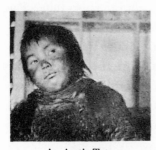

An Arctic Tot.

bert Island, and the glare was still so bright that it hurt my eyes to look at it. I never appreciated the sunlight so much before; involuntarily it made me feel nearer home. The sky was beautifully tinted— pink and blue in the east, light orange in the south, a deep yellow and crimson in the northwest. Fleecy clouds tinged with rose floated overhead, while the air was calm and balmy. How thoroughly I should have enjoyed my walk amid the exquisitely colored surroundings had I known how it fared with my husband on the ice above! Reaching the house at 1.45, I found no tidings of the party, and so watched and waited, until at last a lone figure rounded the mile point. Although I could not see anything beyond a dark spot on the ice moving toward the house, I knew it must be Mr. Peary, for, in spite of his long-forced inactivity

and his broken leg, he still outwalks the boys.   I started out
with Jack, and we soon met.   The party were all right, but
had had a pretty hard time of it.

Thursday, February 18.   A bright, sunny day.   We have
been busy rebuilding the snow entrance which was washed
away by the recent thaw and rain.   This completed, Mr. Peary
got out his " ski " and began coasting down the hill back of the
house.   Astrup and the doctor joined in the sport, and even
the huskies got their sleds and coasted on them.   I spent the
time in taking photographs of the boys, especially in their
grotesque tumbles.

Friday, February 19.   Another cloudy day ; it seems as if
the sun had not yet become accustomed to his new route and
forgets us every other day.   The old couple started for Netchiolumy this morning, and Ikwa went off with his sledge and
our mikkies to bait fox-traps.   After lunch Astrup and the
doctor went on the cliffs to build three cairns from Cape
Cleveland to Three-Mile Valley, expecting to get back by
supper-time.   At six o'clock they had not returned, but we
were not alarmed, and put their supper away for them.   About
seven Ikwa came in, and reported that while passing Cape
Cleveland he had heard the rumbling of a snow-slide down the
steep sides of the cliffs, but it was too dark for him to see anything.   At 9.15 the old couple returned, saying the snow was
too deep for them to travel, and they are therefore going to
stay here for a while.   The truth is, they like it here, and
think they had better let well enough alone.   They said that

in passing Cape Cleveland they heard Jack bark and Dr. Cook
halloo to them.   This, together with Ikwa's story of the snow-
slide and the non-appearance of the boys, made us think that
something might have happened to them, so Mr. Peary and
Gibson started for the Cape at once (about ten P. M.).   When
they reached it they heard Jack whining, crying, and barking
by turns, and on going around the Cape they found quantities
of loose snow evidently lately brought down from the cliffs,
and in the middle of this heap a snow-shoe!   Mr. Peary called
and called, but the only answer received was Jack's cry, nor
would the animal come down.   Mr. Peary at once started
back to Redcliffe on almost a run—Gibson had all he could
do to keep up with him—intending to procure ropes, sledges,
sleeping-bags, alpenstocks, lanterns, etc., and to call out all
the men in the settlement in order to begin at once a close
search of the almost vertical cliffs, covered with ice and snow,
where Jack was, and where he supposed the boys might also
be, perhaps badly bruised and mangled, or overcome by the
cold.   In the meantime, to our great relief, both boys appeared
at Redcliffe, exhausted and hungry.   They said they had
reached Cape Cleveland about 1.30 P. M. and started up the
cliff; it was very steep and seemed unsafe for about one third
of the way, but after that it appeared to be easy climbing.
When, however, they had ascended three hundred feet, progress became increasingly difficult, the course being over round
stones covered with ice, where it was impossible to cut steps.
On looking down they found, to their horror, that it would be

MY FAITHFUL COMPANIONS, "JACK" AND "FRANK."

impossible to return, the cliff being too steep and slippery. Here Astrup dropped a snow-shoe—Ikwa's snow-slide—which he had been using to punch steps in the snow and to scrape places among the icy stones for a foothold. This left them only the one which the doctor was using. Further progress was very slow; they knew that their steps had to be firm, for one misstep would send them to their doom. To add to their difficulty it began to grow dark, about four P. M., when they were not more than half-way up; poor Jack was unable to follow them any longer up the steep, icy wall, and, likewise unable to go down, he began to howl and cry piteously at being left. The howl of a dog under the most favorable circumstances is horrible. To the boys it sounded like their death-knell. They heard the old people pass along the bay, and called to them. Finally they reached the top, and then ran along to Mile Valley above the house and came down it to the bay, in this way missing Mr. Peary.

Sunday, February 21. Yesterday we made an unsuccessful effort to rescue Jack, and this morning the attempt was resumed by Mr. Peary and Dr. Cook. I was to meet them at noon with lunch. About ten o'clock the boys reported a wind-storm down at Cape Cleveland; the snow was driving off the cliffs in thick clouds, and the whole sky became black. The storm, however, did not strike Redcliffe, but passed to the east, and we could see it at work at the head of the bay. Believing it to be over at the Cape, I started on snow-shoes, with shotgun on my shoulder, and with a gripsack containing tea, boiler,

8*

cups, spoons, alcohol-stove and alcohol, potted turkey and biscuits, and sugar and milk.   On turning the first point the wind struck me, but, thinking it was only a squall left by the recent storm, I hastened on as best I could.   Finally I left the path and went inshore, but could not see where I stepped, and was blown down several times.   I relieved myself of the snow-shoes and gun, but was again knocked about by the wind, and had my breath completely taken away by the snow driving in my face.   I finally met Mr. Peary with our good dog Jack, and we reached home late in the afternoon, tired and sore.

Monday, February 22.   Washington's birthday; grandmother's birthday.   Our dinner consisted of venison pie with corn, broiled guillemot breasts and green peas, chocolate, and apple pandowdy.   The day has been cloudy and misty.

Sunday, March 6.   I am recovering from an attack of the grippe.   Tuesday, February 23, after going to bed I had a chill, and all night my back and every bone in my body ached.   In the morning my aches increased and I was in a fever.   Of course Mr. Peary called in the doctor, and between them they have brought me round.   I went out for the first time yesterday, Mr. Peary pushing me on the sledge to the tide-gage, where the sun was shining beautifully.

Tuesday, March 8.   Yesterday was a bright, cold day.   Matt returned from a four days' deer-hunt at the head of the bay, during which he experienced a temperature of from $-40°$ to $-50°$.   Gibson has had everything he possesses put in order for a hunt with Annowkah, in Five-Glacier Valley. He took two reindeer sleeping-bags, his full deerskin suit, a sealskin suit, heavy woolen shirts, stockings *ad libitum*, a heavy pair of blankets, a tarpaulin, and sundry small articles, besides an Eskimo lamp and blubber, which he proposes to keep burning in the igloo all the time.

Tuesday, March 22. The last two weeks have been entirely uneventful, our time having been largely occupied in preparations for various hunting-trips and the great inland journey —the fashioning of experimental clothing, making of sledges, etc. The temperature has been steadily rising, but we have had some sharp reminders of an Arctic winter's force; on the 14th, when the sun shone for the first time on the window of our room, the mercury was still $-35°$. The landscape is now resplendent in its glory, but the beauties of the snow-plain are here wasted on the desert air. Day before yesterday Mr. Peary made a reconnoissance of the ice-cap, traveling about twenty-two miles, and reaching an elevation of 3800 feet; his minimum temperature was $-32°$ as against $-25°$ at Redcliffe. To-morrow he intends to start for Netchiolumy.

Sunday, April 3. The past week has been a long and anxious one for me. Mr. Peary's indisposition last Sunday turned out to be an attack of the grippe, and for two days he was very sick, his fever running up to 103.8. It was accompanied with vomiting, coughing, and violent headache. Tuesday night his temperature went down to normal, and he felt better but weak, and this weakness he fought against with the un-

reasonableness of a child.   Wednesday he said he would start
for Netchiolumy, in spite of my protestations, telling me I was
childish to suppose he did not know what was best for him;
and not until the doctor told him that there was danger of
pneumonia, and that he must take the responsibility if he persisted in going, did he reluctantly yield.   Thursday night his
temperature began to rise again in consequence of over-exer-
tion.   Friday he still fought against lying down and keeping
quiet, and Saturday and Sunday he had a relapse, his fever
reaching 102.2, and leaving him weaker than before.   I have
done nothing but watch over him, and it has kept me busy
day and night.

The weather during the week has been beautiful, and the

One of our Visitors.

sunshine is appreciated by us more and
more every day.

Yesterday, late in the evening, two men
were seen coming toward the house from
the direction of Cape Cleveland.   They
proved to be Kyo and Keshu, the Cape
York dudes.   They said quite a number of
people were in a deserted igloo on Herbert
Island and would be along by and by.   It
seems our former visitor, the widow Klayuh,
whose husband was drowned while har-
pooning an oogzook seal last fall, and who
stopped here with her three children on
her way to Cape York to see her dying

father, has consoled herself by becoming Kyo's wife, and she is
among those who are to come.   This morning both Eskimos
started off to bring their friends, together with their sledges
and dogs, over to Redcliffe.   As Mr. Peary is anxious to get
some dogs, he sent Gibson and Astrup to follow them and see
that they brought all the animals with them.

Monday, April 4.   About two o'clock this morning our
expected visitors arrived, and reported that they had seen
nothing of Gibson and Astrup, nor of Kyo and Keshu.   The
arrivals are Klayuh and her two children—the elder, Tooky,
apparently a young lady (as she has her beau in tow), although
they give her age as only twelve suns; and the younger, a
girl of five or six suns—Tooky's admirer, Kookoo, Klayuh's
stepmother, a widow of three months, with her small child on
her back, and her beau Ahko.   Not knowing that her husband was dead, and in order to say something to her when
she came in my room, I asked her if the man accompanying
her was her husband, when, to my surprise, she burst into
tears and sobbed out that her husband was dead.   I began to
talk in a sympathetic manner, when she suddenly dried her
eyes and interrupted me with, "Utchow, utchow, mikky
suṅgwa Ahko wenia awanga" (wait, wait a little while, and
Ahko will be my husband).   This forenoon another couple
arrived, both rather youthful in appearance, and the woman
quite small; they too had seen nothing of the boys.   Just as
we were through with dinner Astrup came in and said Gibson
was coming with Kyo and Keshu and eight dogs; in about

an hour and a half they arrived.   After dinner I helped Mr.
Peary reload one of his cameras, and in this operation I could
see how nervous he still is.   For the first time since I have
known him he has the blues, and pretty badly at that.   He has
lost confidence in himself, and is harder to nurse than after
his accident on board of the " Kite."   However, he insisted on
photographing and measuring all the newcomers, and this kept
us up until nearly two o'clock — Mr. Peary photographing, the
doctor measuring, and I recording.   I saw that he was very
much exhausted, and I gave him his salt-water sponge-bath
under the blankets, after which he slept well, something he
has not done of late.

Wednesday, April 6.   Yesterday the sun was warm enough
to melt the snow on top of the house, and I put my eiderdown
pillows out for an airing.   To-day has been so lovely that the
women took their sewing on top of the house, where they also
took their babies, stripped them, and placed them on a deerskin, allowing the sun to beat upon them.   The little ones
crowed and seemed to enjoy it hugely.   In company with
Astrup and Annowkah Mr. Peary sledged across to Herbert
Island to get some blubber for Annowkah's family that had
been cached there last summer.   He got back at midnight
and looked very tired, having walked at least twenty-five
miles, but he is in better spirits, and I hope the trip will benefit him in spite of his fatigue.   During his absence I thawed,
scrubbed, cut up, and tried out twenty-five pounds of bacon,
getting twelve pounds of clear fat; I also cut up and tried
out four pounds of toodnoo (venison tallow), which gave me

two and a half pounds of grease. This is to be utilized in the
lunches for the advance party. It took me about eight hours
to do all this.

Saturday, April 9. This morning we found the doctor down
with the grippe. Poor fellow, I am afraid he will have a hard
time of it. The boys have no consideration for the sick, and he
is right out in the noise and turmoil all the time. At eleven A. M.
Mr. Peary started with his six best dogs and Keshu for Herbert Island to bring back some seals cached there for dog-
food. He rode the whole distance over, which, measured by
the odometer, was 14.06 miles. During his absence I worked
on canvas-bags for various instruments and on cording the
sails intended for our sledges. At 11.30 P. M., it being daylight throughout the twenty-four hours, I started to meet Mr.
Peary, but had only walked half a mile when I saw him coming. The day has been, as usual, fine; temperature ranging
from —9° to —22°. We have now a team of ten good dogs,
a very cheering sight for us. Mr. Peary feels confident that he
will get more, and this means assured success on the inland ice.

CHAPTER XI

SLEDGE JOURNEY INTO INGLEFIELD GULF

The Start from Redcliffe — Our Team — Temporary Village on Northumberland Island — A Crazy Woman — A Never-to-be-forgotten Night in a Native Snow-igloo — From the Snow-village to Keati — Across Whale Sound to Netchiolumy — An Eskimo Metropolis — Aged Dames — From Netchiolumy to Ittiblu — Midnight Glories — The Solitary Habitation at Ittiblu and its Inhabitants — My Coldest Sleep in Greenland — Nauyahleah, the Ancient Gossip — A Native Graveyard — From Ittiblu to the Head of Inglefield Gulf — Meeting with a Traveling-party.

Monday, April 18. Having completed our arrangements for a week's exploration of Inglefield Gulf, we started from Redcliffe about noon with the large dog-sledge, drawn by six dogs and driven by Kyo.

The day was very bright, and the sun shone warm all the time. The traveling as far as Cape Cleveland was good, but then it began to grow heavy, and before we had gone half-way across there were places where the dogs sank in to their bellies and almost swam, while we sank down to our knees in a semi-slush; the sledges, however, went along nicely. Fortunately, there were only a few such places, and as we got near the west end of Herbert Island the ice became smoother and harder, and the dogs sped along, two of us riding at a time, and sometimes all three.

Our sledge reached the west end of Herbert Island at eight o'clock, and two hours later, having crossed over to Northumberland Island, we came upon a cantonment of four snow-igloos. These were occupied by families from different settlements, who congregated here to be near a patch of open water a short distance off, where they caught seal. The largest snow-igloo was occupied by Tahtara, his wife, his father and mother, and some small children. This was put at our disposal; another was occupied by Ikwa and family, together with Kyoshu and his son, while Myah and his wife were accommodated in a third. The mistress of the remaining igloo was making an awful noise and trying to come out of her habitation, while a man was holding her back and talking to her, but she screamed and struggled so long as we remained where she could see us. I asked Mané what was the nature of the trouble, and she told me that the woman was pi-bloc-to (mad).

As the wind was blowing fiercely and the air was thick with drifting snow, Mr. Peary urged me to come into the igloo, which I did, rather to please him than to get out of the storm. Now as long as I have been in this country I have never entered an Eskimo hut; hearing about the filth and vermin was quite enough for me. But Mr. Peary said the snow-house was much cleaner, etc., etc., and seeing that it really made him uncomfortable to have me stay outside, I yielded. Can I ever describe it? First I crawled through a hole and along a passage, about six feet, on my hands and knees; this was

level with the snow outside. Then I came to a hole at the
end of the passage and in the top of it, which seemed hardly
large enough for me to get my head through, and through
which I could see numberless legs. Mr. Peary called for me
to come, so the legs moved to one side and I wedged myself
into the aperture and climbed into a circular place about five
feet high, the floor of which, all of snow, was about two feet
higher than that of the tunnel. A platform one and a half
feet above this floor, and perhaps four feet wide in the middle
and two and a half feet at the sides, ran all around the walls
of the igloo, except that part in which the aperture or door
came up in the floor. The middle of this platform for about
five feet was the bed, and it was covered with two or three
tooktoo skins, which almost crawled away, they were so very
much alive. On this bed sat Tahtara's mother, tailor-fashion,
with a child on her back; another woman, younger by far,
and rather pretty, his wife; and two children, about six and
eight years old; and on the edge, with his feet resting on a
chunk of walrus, from which some hungry ones helped themselves whenever they wanted to, regardless of the fact that a
number of feet had been wiped on it, and that it was not only
frozen solid but perfectly raw, sat Tahtara himself, smiling and
saying, "Yess, yess," to everything that Mr. Peary said to
him. Mr. Peary had also taken a seat on the edge of this
bed, and the women immediately made room for me between
them; but this was more than I could submit to, so, excusing
myself by saying that my clothing was wet from the drifting

snow and that I could not think of getting their bedding wet, I sat down, not without a shiver, on the edge beside Mr. Peary, selfishly keeping him between the half-naked women and myself.

The sides of this platform on either side of the doorway were devoted to two ikkimers (stoves), one of which was tended by Tahtara's mother and the other by his wife. These stoves were very large and filled with chunks of blubber; over each hung a pan, made of soapstone, containing snow and water, and above these pans were racks or crates, fastened very securely, on which the inmates flung their wet kamiks, stockings, mittens, and birdskin shirts. The drippings of dirt, water, and insects fell invariably into the drinking-water. I say " drinking-water "; they have no water for any other purpose. Mr. Peary had put our Florence oil-stove on the side platform and was heating water for our tea. Fortunately our teapot had a cover on it, which I made my business to keep closed.

Besides the persons mentioned there were always as many husky visitors as could possibly pack in without standing on one another. These took turns with those unable to get in, so that after one had been in a while and gazed at the circus, he would lower himself through the trap and make way for a successor among the many crouching in the passageway behind him. This was kept up throughout the night. Of course the addition of our stove, together with the visitors, brought the temperature up rapidly, and to my dismay the

Eskimo ladies belonging to the house took off all of their clothing except their necklaces of sinishaw, just as unconcernedly as though no one were present.

The odor of the place was indescribable. Our stove did not work properly and gave forth a pungent smell of kerosene; the blubber in the other stoves sizzled and sometimes smoked; and the huskies—well, suffice it to say that was a decidedly unpleasant atmosphere in which I spent the night.

I soon found that if I kept my feet on the floor they would freeze, and the only way I could keep them off the floor was to draw up my knees and rest the side of one foot on the edge of the platform and place the other upon it. In this way, and leaning on my elbow, I sat from ten at night until ten in the morning, dressed just as I was on the sledge. I made the best of the situation, and pretended to Mr. Peary that it was quite a lark.

Mr. Peary went out to look after the dogs several times during the night, and each time reported that the wind was still blowing fiercely and the snow drifting. In the morning the wind had subsided somewhat, and after coffee the dogs were hitched, and we resumed our journey, heading for Keati.

After traveling about an hour we came upon a single stone igloo, which proved to be Nipzangwa's; he and his father, old Kulutunah, immediately came out to meet us. We reached Keati, the inhabitants of which had been apprised in advance of our coming by special messenger, about noon, and an hour later, reinforced with additional dogs, started across the Sound

THE INHABITANTS OF "SNOW VILLAGE," NORTHUMBERLAND ISLAND.

for the settlement on Barden Bay (Netchiolumy).   Ikwa followed with his dogs and sledge.   The traveling was fine, and the dogs took our sledge, with all three of us riding, along at a trot all the way.   We arrived at our destination about six P. M., the odometer registering 14.4 miles from Keati.

Here we found a great many natives, probably sixty, most of whom we had already seen at Redcliffe during the winter.

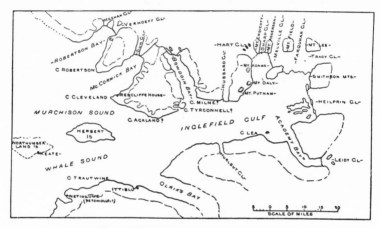

Map of Whale Sound and Inglefield Gulf.

In addition to the regular inhabitants of the place there were a half-dozen families from Cape York and its vicinity, who were stopping in snow-igloos on their way home from Redcliffe. The winter is their visiting time, and only during this season do the inhabitants of one place see those of another; they travel for miles and miles over the ice, some with dogs and

9

some without, but there is invariably at least one sledge with
every party.   This year the travel has been unusually brisk,
owing to the American settlement, which all were anxious to
visit.   Where a family has a sledge and two or three dogs,
they load it with a piece of raw walrus or seal (enough to last
them from one village to the next), anything and everything
that can be scraped together for trade, one or two deerskins
for bedding, and the smallest child that has outgrown the
mother's hood.   The rest of the family then take turns in riding, one at a time, while two push the sledge.

On our arrival at the igloos we were immediately surrounded by the natives; two very old women in particular
were led to me, and one of them, putting her face close to
mine—much closer than I relished—scrutinized me carefully
from head to foot, and then said slowly, " Uwanga sukinuts
amissuare, koona immartu ibly takoo nahme," which means,
" I have lived a great many suns, but have never seen anything like you."

We had brought our things up to the igloos and intended
to get our supper on the hill, but the native odor, together
with that of *passé* pussy (seal) and awick (walrus) lying about,
was too strong, and I suggested that we return to the sledge.
The two old women who first greeted us, despite the fact that
they could not walk alone, were determined to accompany us,
and they were helped down the hill to the sledge.   They
looked as old and feeble as women at home do between eighty
and eighty-five.   Never having seen such a sight, they could

not let the chance go by, even at the expense of their little strength. Not being able to carry everything in one trip, I went back for the rest, preferring this to staying with the sledge, where the natives were now swarming, and wanting to handle everything they saw. When I came to the igloos again, Annowee, a Cape York woman, who had lately been to Redcliffe, began to beg me not to go down, but to have Mr. Peary come up to her; she had " ah-ah " (pain) in her knees and could not possibly make the descent. She wanted to see us as long as she could, as she would never see our like again. All this time she was not only talking loudly, but clutching at my arm whenever I turned to go, and when I said, " Utchow, utchow, wanga tigalay " (just wait, I am coming back), she said, " Peeuk," but did not want me to take the things down for fear I should not come back. The other women now closed about me, and all begged me to stay. Mr. Peary, who remained with the sledge, was somewhat disturbed by my position, but it was all done in kindly feeling. In spite of the fact that Annowee " could not come down," she was at the sledge almost as soon as I was.

We took our supper, after which we bartered for tanned oogzook-sinishaw (seal-thong), sealskins, bearskin trousers, and two dogs. Old Ahnahna gave me a scolding for the benefit of her companions because I would not give her a needle; she said Mr. Peary was " peudiochsoa " (very good) but " Mittie " Peary was " peeuk nahme "—that I used to give her needles, but now I would not do it, etc. While saying this she was laughing all the time, and when I gave her a cup of tea and a cracker she changed her opinion of me at once.

Mr. Peary walked to the Tyndall Glacier and took photos of it, and of the village and the natives. Kyo then hitched up the dogs, we said good-by all around, Ikwa included, and at eight o'clock left for Ittiblu.

To show how sharp these semi-savages are, I may mention the following incident: On the way from Keati to Netchiolu-my we dropped at different times three snow-shoes from our sledge, but seeing Ikwa behind us pick them up, we did not stop for them. On reaching Netchiolumy he brought them to us, and said they were fine for us, were they not? We said yes. "Well," he said, "if I had not picked them up you would not have them, and as my eyes hurt me very much, and I see you have them to spare, you should give me a pair of smoked glasses." I thought so too, and he got what he asked for.

We had the perfection of traveling. The surface of Whale Sound was just rough enough to prevent it from being slip-pery, and yet so smooth that the sledge went along as if it were running on a track.

Mr. Peary, Kyo, the driver, and myself were all three seated upon the sledge, which in addition was heavily laden with our sleeping-bags, equipment, provisions, etc., and yet the nine handsome creatures, picked dogs of the tribe, who were pull-ing us, immediately broke into a run, and, with tails waving like plumes over their backs, kept up a brisk gait until we reached Ittiblu at two o'clock in the morning; the odometer registered 21.94 miles.   The night was a beautiful one.   The sun shone brightly until near midnight, when it went down like a ball of fire, tinging the sky with crimson, purple, and yellow lights, which gradually faded out and left a dull grayish blue, which in turn changed to a gray just dark enough to show us the numberless stars that studded the firmament. When we reached Ittiblu the sun came up from behind the dark cliffs of the eastern shore of Inglefield Gulf.   We had been traveling sixteen hours, and were pretty well tired out. Our dogs, too, were glad to have a meal and rest.

We immediately set to work to build a snow-igloo of our own, on the icy floor of which we placed our sleeping-bags and everything that we did not wish handled by the inhabitants of the settlement. While still at work on this we were visited by two residents, Panikpah, a former visitor at Redcliffe, and Koomenahpik, his father;

Our Snow-igloo.

they showed a true native hospitality by asking us to share the comforts of their igloo—an invitation, however, which we politely declined.

Our igloo proved icy cold, and I shall never forget the

9*

difference of temperature between inside and outside. It was just like going from a cellar into a temperature of 90°, and we resolved that unless it was storming we would in future sleep without shelter.   Among our breakfast callers was the wife of Koomenahpik, Nauyahleah, the most comical old soul I had yet seen. She evidently felt it her duty to entertain me, and began to tell me all about herself and her family; she let me know that I had already seen one of her sons at Redcliffe, whose name is Tawanah, and who lives still farther up Inglefield Gulf; he had stopped at Ittiblu, she said, on his return from the Peary igloo, and told her what a large koona Peary's koona was, and how white her skin was, and that her hair was as long as she could stretch with her arms. She followed us wherever we went, and chatted incessantly — whether we were taking photographs or making observations for latitude and time, it made no difference to her. If we did not answer her she would sing at the top of her voice for a few minutes, and then chatter again. She showed us a number of graves, which are nothing but mounds of stones piled on the dead bodies, and told us who lay beneath the rocks.

At eight in the evening we left Ittiblu, with four additional dogs obtained from Panikpah. All night long we dashed on over the smooth surface of Whale Sound, except where we passed Academy Bay. Here from one cape to the other the snow was soft and several inches deep. Again the sun only left us for a short time, and in spite of a temperature of −35°, the ride was a delightful one.

About two A. M. we were abreast of another beautiful glacier, a great river of ice slowly making its way from the eternal inland ice to the sea. The smooth and even appearance of all the glaciers, Mr. Peary told me, was due to the blanket of snow which covered them.

It took us about an hour to pass the face of the ice-sheet, which in places towered above us to a height of one hundred feet and more. As we rounded the southwest corner Kyo sang out, "Inuits, Inuits," and, looking ahead, we saw an Eskimo snow-igloo built up against the rocks on the shore. Scattered about on the ice-foot lay about a dozen seals, some whole, and some partially cut up; three or four young white seals, a number of sealskins, a large sledge and a small toy-sledge patterned exactly like the large one, and coils of seal-skin and walrus lines. In the "tochsoo," or entrance to the igloo, was tied a young dog, who had no idea of awakening his master, for he only looked at us and gave no sound.

In response to Kyo's shouts a man came slowly crawling out, rubbing his eyes, and showing every evidence of having been suddenly awakened out of a sound sleep. This proved to be Kudlah, a young native whose home was at the head of Inglefield Gulf, and who on a visit to Redcliffe during the winter had been nicknamed by our boys " Misfortune." Kud-lah had a hang-dog sort of expression. We were told that a woman would only live with him a year and then leave him, it being the privilege of the Eskimo maiden to return to her parents' roof at the end of a year, provided there is no family, if she finds that she has made a mistake. "Misfortune" had grown very fond of the "kabloonah's kapah" (white man's food), especially coffee and crackers, during his visit at Redcliffe, and he now came right to our sledge and asked if we had no "kapah" for him. He told us that he, with his wife, and Tawanah with his wife, a son twelve years of age, and three smaller children, were on their way to Redcliffe. They had left their home, Nunatochsoah, at the head of Inglefield Gulf, two days before, and had walked all day and until midnight, when they built the snow-house and camped. The women and children being very tired, and seal-holes, whence young seals are procured, being plentiful in this neighborhood, they decided to rest a few days and hunt seal. I asked him where they found the pretty little white creatures, and he told me that the mother seal crawls out on the ice through the cracks and hollows out a place for herself under the snow, not disturbing the surface at all, except perhaps by raising it a little, and thus giving it the appearance of a snow-drift or mound. Here she gives birth to her young, and stays with them until they are old enough to take to the water, leaving them only long enough to get food for herself.

To me these mounds did not seem different in appearance from the ordinary snow-mound, but the trained eye of the native immediately distinguishes the "pussy igloo" (seal-house); he walks softly up to it, and puts his ear close to the snow and listens. If he hears any sign of life he jumps on the mound as hard as he can, until it caves in, and then, with a

SLEDGING INTO INGLEFIELD GULF.

kick in the head, he dispatches the young one.   Then he lies in wait for the mother seal to return to her young, when she is promptly harpooned.

While Kudlah was entertaining us, Tawanah and the two women came out of the igloo.   The latter were very much interested in me, and wanted to know if there were any more women like me at Redcliffe.   When told that there were not, but that they were plentiful in the American country, they asked, "Are they all so tall, and so white, and have they all such long hair?   We never have seen women like you."

Our driver had been refreshing himself with seal and blubber, and Mr. Peary now called to him to untangle the dogs, as we wished to continue our journey.   This he did not like, and said the people were all gone, and there was no use in going any farther up the gulf.   The snow, he said, was very deep, and the dogs would not be able to pull the load; but Mr. Peary was firm in his decision to push on to the head of the gulf, if possible, in order to complete his surveys.   Accordingly, at four A. M. we started again, and to our surprise Kudlah and Tawanah accompanied us.   When questioned as to their destination, Tawanah said they had a lot of sealskins and young seals at Nanatochsuahmy which he wanted to give Mr. Peary, and they were going as far as his igloo with us.

In about three hours we came to a small island, and here we pitched camp.   After a hearty supper of Boston baked beans, corned beef, and stewed tomatoes, with tea and crackers, we turned in, and what a delightful sleep we had!   The sun shone warm, and that peculiar stillness which is found only in the Arctic regions was conducive to long sleep.

After supper we explored the little island and found the plateau covered with the tracks of deer and ptarmigan, but we could descry no living creature. The view from the summit was very fine. We could see down the sound as far as

Herbert Island, and almost up to the head of Inglefield Gulf; on the right the eye took in the greater part of Academy Bay, and on the left in the distance towered Mts. Putnam, Daly, and Adams.

Arriving at Nunatochsoah, we spent about an hour in skirmishing about the

Mount Daly.

place, Tawanah taking us to various caches containing sealskins, both tanned and untanned, and two caches containing young seals, about twenty-two in all. Kudlah, too, had a few seals and skins, and both men were anxious to barter their possessions with Mr. Peary for a knife and a saw.

CHAPTER XII

THE SLEDGE JOURNEY — (Continued)

From Tawanah's Igloo to the Great Heilprin Glacier — The Little Matterhorn —
A Wet Night — Ptarmigan Island — "As the Crow flies" for the Eastern Bastion
of Herbert Island — A Nap in the Sunshine — Back at Redcliffe — A Busy Week
of Preparation for the Start on the Inland Ice — Canine Rivals.

We unloaded our sledge, and, with Kudlah as our driver, continued the exploration of Inglefield Gulf to its head. In spite of Kudlah's having spent the entire time at Tawanah's in eating seal, we had scarcely traveled a mile before he said he was hungry for American kapah. When told it was not yet time, he turned his attention to the dogs again, but soon we saw that the dogs were having a go-as-you-please time, and on looking to the driver for the reason we found him sitting bolt-upright and fast asleep. We woke him, and to keep him awake I gave him some crackers to eat. They had the desired effect as long as they lasted, but as soon as they had disappeared off he went to sleep again, and I came to the conclusion that they acted more as a narcotic than a stimulant, and discontinued them.

Just before reaching the head of this great gulf we came to a nunatak in one of the numerous glaciers, shaped like the

Swiss Matterhorn, and we named it the Little Matterhorn. We were in an Alpine landscape, but the more striking features of the European ice-covered mountains were here brought out in increased intensity. Arrived at the head of the gulf, we were confronted by one of the grandest glaciers that we had yet seen.

Never shall I forget my impressions, as, on this bracing April day, with the thermometer from 30° to 35° below zero, Mr. Peary and I, shod with snow-shoes, climbed over the deep-drifted snow to the summit of a black rock, destined in a few years to be engulfed by the resistless flow of the glacier, and from this elevated point looked out across the mighty stream of ice to the opposite shore, so distant as to be indistinct, even in the brilliant spring sunshine that was lighting all the scene. Looking up the glacier, the vast ice river disappeared in the serene and silent heights of the ice-cap. To think that this great white, apparently lifeless, expanse, stretching almost beyond the reach of the eye, is yet the embodiment of one of the mightiest forces of nature, a force against which only the iron ribs of mother-earth herself can offer resistance! As we stood there silent, a block of ice larger than many a pretentious house, yet but an atom compared with the glacier itself, pushed from its balance by the imperceptible but constant movement of the glacier, fell with a crash from the glacier face, sending the echoes flying along the ice-cliffs, crushing through the thick bay ice, and bringing the dogs, far below us, to their feet with startled yelps.

The glacier, which forms much of the eastern wall of Ingle-field Gulf, has a frontage of about ten miles, and is the largest of the series of giant glaciers in which are here concentrated the energies of the ice-cap. North of it lie the Smithson Mountains, and farther beyond, a vast congeries of ice-streams which circle westward and define the northern head of the gulf. To the eastern sheet, upon whose bosom no human being had ever stepped, and on whose beauty and grandeur no white person had ever gazed, we gave the name of Heilprin Glacier, in honor of Prof. Angelo Heilprin, of the Academy of Natural Sciences of Philadelphia.

On the upward voyage to Greenland we had passed num-bers of glaciers, beginning with the great Frederikshaab ice-stream. I had seen the distant gleaming of the Jacobshavn Glacier, and after passing Upernavik we were never without a glacier in sight, and yet it was not until September, when Mr. Peary was able to get out in the boat, and we went to the head of McCormick Bay to see the inland ice-party off, that I came in actual contact with one of these streams of ice. About eight miles above Redcliffe, on the same side of the bay, there is a hanging glacier, which has peered at us past the shore cliffs ever since we entered McCormick Bay. This glacier is supported upon a great pile of gravel, looking like a railway fill, which gives it the appearance of being upon stilts. It was a peculiar experience to see the red-brown rocks and cliffs glowing in the sun, and this great vertical wall of blue ice standing out beyond them, with little streams of water trickling

down from it, and occasionally fragments of ice breaking away
and dashing down with a muffled, metallic sound; and more
than this, to find the ever-constant friend, the Arctic poppy,
growing actually beneath the overhanging walls of the glacier.
The great glaciers, too, that surround Tooktoo Valley, with its
green meadows and glistening lakes, will always remain with
me an exquisite recollection.

Returning to our sledge, we made a direct line for our camp,
which was reached after an absence of ten hours.

Wearied with our journey, we immediately prepared to rest,
and selected a sheltered nook on the sea ice, where the snow
was several inches deep, and where we were protected from
the light breeze which blows almost constantly by a huge but-
tress of ice, part of the ice-foot. The memory of the delightful
sleep of the night before, when we lay right out in the sunshine, helped me to hurry the sleeping-bags into place and
crawl into mine without losing much time.

Tawanah came to me and asked if I would not like to have
my kamiks and stockings put up on the rocks where the sun
could shine on them and dry out what little moisture they
might contain, and I told him to take them away. In what
seemed to me only a few minutes, but what was actually four
hours, I was awakened by some one grasping both sides of
my sleeping-bag, evidently trying to stand it and its contents
on end. The words "Don't roll over; try to stand up as
quickly as you can; the tide has risen above the ice," rang in
my ears. On looking about me I saw that I had been lying
in about six inches of water and peacefully sleeping.

Fortunately I had a sealskin cover over my deerskin bag, and the water had not penetrated it; therefore my deerskin knickerbockers and flannel wrapper, which I always take off after I have pulled myself down in the bag, fold and place under me, were perfectly dry. My poor husband did not fare so well. He had folded his trousers, kamiks, and stockings and placed them under his head as a pillow, and of course they were soaking wet. Not having a cover to his sleeping-bag, the water had soaked through, and it was this that had wakened him.

After a time we managed to dry out, and, continuing our journey, reached our little island at midnight. As we approached the island numbers of ptarmigan were seen flying about the rocks, a circumstance which determined us to name the spot Ptarmigan Island. We secured a few of these beautiful, snow-white birds, and, after taking observations for position, proceeded on our course to Tawanah's igloo, which we reached shortly after four A. M.

While preparing the morning meal, I was the center of an admiring circle. Men, women, and children formed a perfect ring about me. Never had they seen such a stove, and never such cooking. They chattered incessantly, and plied me with so many questions that I began to despair of getting anything to eat. Finally I gave each a tin of coffee and some crackers, and this kept them busy long enough for me to eat my meal, and we then turned in.

We awoke about four o'clock in the afternoon, and at once began our exploration of the surrounding cliffs and the neighboring glacier, which Mr. Peary considered one of the first magnitude, and named, after the distinguished secretary of the American Geographical Society, the Hurlbut Glacier. It was nine o'clock before we were through with exploring, photographing, and making observations, and then we made a dash for the east end of Herbert Island.

Mr. Peary laid our course down the center of the gulf, and we were beginning to calculate the time when we should reach Redcliffe, when suddenly we encountered deep, soft snow, through which the dogs could not pull the loaded sledge with any of us seated upon it. There was nothing left for us but to get off and walk, or rather wade through the snow. After a few hours of this tiring work the dogs refused to go farther, and it was only with special coaxing and driving that any progress was made. When at last we reached Herbert Island we were almost as glad as the dogs to be able to rest. Redcliffe was still fifteen miles distant.

Mr. Peary and I spread our sleeping-bags down on the snow out in the brilliant sunshine, and lay down on them for a nap. We had not been asleep long when I awoke and found that Mr. Peary had arisen and was walking rapidly in the direction of the ice-foot. He was following an Eskimo who had shouldered a rifle, and my first impression was that the native had taken one of our own rifles from the sledge and was making off with it.

At Kyo's call the retreating figure stopped short and turned back. He came directly to us, and we recognized him as

AN APRIL JOURNEY.

Tahtara, the man at whose snow-igloo I had spent such a memorable night.   He had been at Redcliffe, and was now out on a seal-hunt, with a companion, named Kulutingwah, who presently came dashing round with two fine-looking dogs and one of our sledges.

These dogs were the most affectionate Eskimo dogs we had yet seen, and by far the prettiest.   They were large, powerful-looking animals, that dragged the sledge with three natives upon it through the soft snow as easily as if they had no load at all.   They were the first dogs we had seen who were trained to obey their master's words without the aid of the whip. When Kulutingwah left his sledge-team he did not have to turn the sledge over and stick the upstanders into the snow to keep the dogs from running away, but simply told them to stay there, and with a low, deep growl they would stretch themselves upon the snow and remain perfectly quiet until his return, in spite of the tempting pieces of seal meat which might be lying around in their vicinity.

After restowing our sledges we started homeward.   Our dogs, like horses at home, seemed to smell the stable, and broke into a brisk trot, which they kept up until we reached Redcliffe, at nine in the evening, Sunday, April 24.

Dr. Cook, who had been left in charge, had done good work during our absence of a week.   Quite a number of natives from Netchiolumy, Keati, and the snow village had arrived, and among them an unusual number of lady visitors, all willing to sew for the "Americans" for the small consideration of

10

a couple of needles. The doctor had set them to work on kamiks, fur mittens, fur stockings, and fur trousers, and they had worked like beavers all the week, while the men had put in their time hunting, and a goodly number of seals were added to the store of dog-meat.

We were now in possession of twenty-two good dogs, the pick of all the dogs in the tribe, and Mr. Peary felt that the success of his long sledge journey was assured. Every pack of Eskimo dogs has its leader. If a new dog is added to the

Musical Dogs.

pack a fight takes place at once between him and the leader to determine his position in the team. Now, up to this time a great white shaggy brute, from Cape York, whom we called Lion, on account of his gray mane, had been the canine king of Redcliffe. With the arrival of Kulutingwah's fine dogs there came a change. Lion and his first lieutenant, a dog marked very much like himself, at once charged upon the new-comers, evidently expecting to thrash them into subjection as easily as had been done in the case of the other dogs, but he, for once, was doomed to disappointment; although the fight raged fierce and long, poor Lion was vanquished, and forced to resign his position as king in favor of the larger of the new-comers, whom we called "Naleyah" (chief).

CHAPTER XIII

OFF FOR THE INLAND ICE

The First Detachment of the Inland Ice-party leaves Redcliffe — Departure of the Leader of the Expedition — Rest after the Excitement — Arrival of the Ravens — Return of Gibson and Matt — Gloomy Weather — Daily Incidents at Redcliffe — Spring Arrivals of Eskimos — Eskimos imprisoned in their Igloos by a May Snow-storm — The First Little Auks — Open Water off Cape Cleveland — Harbingers of Summer — Myriads of Auks and Seals — Snow-buntings — Green Grass and Flies — Kyo, the Angekok.

Saturday, April 30. The past week has been one of hustle and bustle. The overland ice journey has been uppermost in our minds and actions, and this morning the real start was made. All the boys except Verhoeff, with the dogs and five natives, left with three loaded sledges for the head of the bay, whither several loads of provisions had already been transported. Mr. Peary is to follow in a few days.

Wednesday, May 4. At 8.30 P. M. yesterday, Mr. Peary with Matt, who had returned for additional equipment, started for the head of the bay to join Gibson, Astrup, and Dr. Cook, who have been there since Saturday. I watched him out of sight, and then returned to the house, where Mr. Verhoeff and I will keep bachelor and maid's hall. For three full months I shall be without my husband—a year of anxiety and worry to me.   It has been arranged to have two of the boys accompany the expedition, merely as a " supporting-party," and their farthest point will probably be the Humboldt Glacier; I can therefore expect news from the interior in three weeks or less. The last ten days have been one continuous rush for me, and

Preparing for the Start.

part of the time I hardly knew where I was.   After I am rested I shall begin a thorough overhauling of everything, and get things ready for packing.   As I write, 11.45 P. M., the sun is shining, and as I think Mr. Peary will begin his march to-night, I hope this morning's snow-storm has cleared the weather for some time to come. Strange coincidence: just six years ago I bade Mr. Peary good-by as he started on his first Greenland trip. May it be a good omen, and he return as successful as he did then!

Saturday, May 7. The weather continues alternately dreary and pleasant, but the approach of springtime is unmistakable. Already the ravens have arrived, and moderate thaws have begun to loosen our covering of snow and ice. Shortly after six this morning I was awakened by hearing one of the huskies cry, " My tigalay, my tigalay " (Matt has returned), and in a minute later Matt and Gibson came in. The former had returned on account of a frozen heel, while Gibson came back for additional alcohol. In a note to me Mr. Peary stated that he had met with a severe obstacle in the way of heavy snow and steep up-grades, and therefore had not made the distance that he had hoped to cover in a week's time.

Sunday, May 8. At last it seems to have cleared, but still the head of the bay is enveloped in mist. Gibson left us again yesterday, and he is probably with his party this evening. The thermometer is steadily rising, and with a temperature to-day of 28° everything has been dripping. I got all the snow off the roof of the house and the canvas-covered annex on the west side, as water had begun running down between the tarred paper.

Tuesday, May 10. All night the wind blew a gale from the east and northeast, and all day the snow has been flying

10*

in clouds so thick that at times we could not see the tide-gage, a hundred yards distant. My thoughts have been continually with the little party on the ice. I know who will have the worst time, who will have to look out for everything, and it worries me because I know he is not as well as he ought to be. Everything around Redcliffe is hidden in the snow-drifts, and the snow has been coming in under the canvas until we have three feet of it in front of our door inside the in-closure, in spite of Matt's blocking all the openings in the walls. With Matt's help the range and lockers were moved out of my room to-day, and we found the wall and floor cov-ered with ice. I knocked off as much as I could, and removed the cardboard from the floor, and to-night the blanket and carpet at that end of the room have thawed and are dripping wet. This evening Kyo wanted to know if we would permit him to go with us beyond Cape York, to where the other Eskimos live (Upernavik, or Disko). I told him he could; then he wanted to know if I would draw a map of Greenland, and mark our route upon it. He seemed to understand, and was pleased to know that he could go.

Wednesday, May 11. A beautiful day. The drifts are hard as marble. Matt shoveled the snow out of the entrance, and we once more opened our windows. The drip from the roof has forced us to remove all the snow and ice, and we are thus recovering our non-wintry appearance.

Friday, May 13. Contrary to all expectations, last night and to-day have been warm and bright. All the huskies gathered on our roof, which is dry and retains the sun's heat. Noyah, the baby, rolled about entirely naked in a temperature of 22°, except for a cap, which was nothing more or less than the toe of one of Mr. Peary's cast-off blue socks. Verhoeff, who has made a tour to one of the neighboring icebergs, reports that the snow has been swept from the ice in the middle of the bay, and that the ice has commenced to melt.

Saturday, May 21. The past week has seen our home again converted into an Eskimo encampment. There have been numerous arrivals of old and new faces, representing all conditions of age from the tiniest baby to Tahtara's mother. The simple folk have come as heralds of the approaching spring, some to stay and others to proceed farther. They report the return of the little auk at Keati. Yesterday and to-day have been wild, stormy days, the wind blowing a gale from the southeast nearly all the time, and when it was not actually snowing the snow was flying so furiously that it was all but impossible to face it. The two Eskimo families in the snow-igloos experienced much discomfort, and this morning Kyo called for Matt to dig him out. The snow had drifted in the entrance to his igloo until it had filled and piled up higher than the house, and he had had great difficulty in keeping an air-hole open during the night.

Monday, May 23. A beautiful day. I hoisted a new flag on Redcliffe House in honor of my sister Mayde's birthday. Yesterday was the anniversary of my own birth, the first of my life when I did not receive a birthday wish from my dear mother, and the first which I spent without receiving a loving greeting from some dear one. I was obliged to go through the routine formality of setting out the wine, but I felt neither like eating nor drinking. Yesterday morning the first little auks were seen flying over Redcliffe House, some in the direction of the head of the bay, others in the opposite direction.

Kyo, Matt, and I indulged in a little target-shooting to-day with my revolver. We put up a tin at forty feet distance and fired six shots each. In the first round Matt scored nothing, Kyo hit the target 3 times, while I hit it 5 times. I then stepped out, and Matt and Kyo tried again, the former scoring 5 and the latter 4.

Thursday, May 26. A perfect day, clear, calm, and warm. Nearly four weeks have elapsed since Mr. Peary left me, and yet no news. For a full week, day by day, I have been expecting the supporting-party, and am now nearly desperate. Being in no mood for writing, reading, or sewing, I called Jack and started for Cape Cleveland, where open water had been reported. For a quarter of a mile before reaching the Cape I sank into water almost to my boot-tops, but I felt fully repaid for my trouble by the beautiful sight which met my gaze. The water, of deepest blue and clear as crystal, sparkled and danced in the sunlight, as if it were overjoyed to have broken loose from its long imprisonment, and once more have the countless birds sporting on its bosom. The water and the air above it were at times black with birds, the majority being little auks. There was, however, a goodly sprinkling of black guillemots and gulls. I also saw a pair of eider-ducks. I watched this scene for some time. Two stately, massive bergs in the center of the pool of dancing water imparted grandeur to the picture—now glistening with the dazzling white of marble, and a moment later black with the myriads of feathered creatures that had settled on them. The sight of the water made me feel more homesick than ever, so I continued my walk around the Cape. At every step I broke into the snow nearly to my hips, and sometimes there was water under it. I saw four pairs of snow-buntings chirping and flitting about among the rocks and patches of grass where the snow had disappeared. They were evidently getting acquainted with each other, and looking for a place in which to make their home. Almost half way between the trap-dyke and Three-Mile Valley I came upon the place where Kulutunah had formerly had his tupic, and where he had left nearly one half of a last summer's seal lying exposed on the ice. About this had gathered hundreds upon hundreds of flies, some large and some small, the first I have seen since leaving Upernavik, I think. I brought some back as specimens. The air was filled with the chirping of birds, the buzz of flies, the drip, drip, drip of the snow and ice everywhere about, and the odor of decaying seal. On my return I climbed over the Cape in preference to rounding it, as I had seen large pieces of ice break off and float out into the dark water. From my elevated position the surface of the ice around and beyond the water looked as if it had had its face badly freckled, so covered was it with black specks; each speck represented a seal taking his sun-bath. Yet it is very difficult for the natives to catch these creatures, as the ice is rotten and will not bear their weight.

On reaching Redcliffe House I saw Kyo dressed in a pair of woven trousers, a blue flannel shirt, and a pair of suspenders given him by Matt, and Mr. Peary's old gray felt hat, which I gave him a day or two ago, and which he hesitated to take, because, he said, it was not mine to give, and Mr. Peary would say on his return, "Ibly tiglipo, ibly peeuk nahme" (you steal, you are no good). He looked precisely like an Indian as he stood there, busy putting up his tent on the brow of the hill directly back of the house. This place has been free of snow for some time and is perfectly dry, while his igloo, as well as the other two, is constantly wet from the melting snow. He is filled with the idea of going to America. Every night he comes for a magazine to look at after he has gone to bed, as he has seen some of the boys do. He says Mr. Peary will be his "athata" (father) and Missy Peary his "ahnahna" (mother) on the ship, and when he gets to America he will learn how to read, and then he won't have to select books with pictures. Whatever he sees he wants to know if he will see it made in America. He tells me that he is an "angekok" (doctor), and that he always cures the people. They never die where he is, and he can make them do just as he chooses. His wife does not seem to care to go to

America, so for the last few days he has borrowed two or three magazines to take into his igloo, where for three or four hours at a stretch he has sat with his wife in front of him and the book between them, swaying himself from side to side, and singing a monotonous sort of tune at the top of his voice. In this way, the other natives assure me, he works a spell over her, and she willingly consents to go with him.

Cape Cleveland.

CHAPTER XIV

WEARY DAYS OF WAITING

Anxious Fears for the Inland Ice-party — A " Red-Letter " Day — Return of
the Supporting-party with Good News — First Flowers — Job's Comforters
among the Huskies — An Attack of Homesickness — The Snow disappearing
— My Confidante, the Brook — The Eider-ducks return — I stand my Watch
with the Others — Matt crippled by a Frosted Heel — We are reduced to a Seal
Diet — A July Snow-storm — Influx of Natives — Open Water reaches Redcliffe
— Matt overhears a Native Plot to kill us.

Monday, May 30. We had a great excitement about 8.30
this evening. A black spot was seen out in the sound beyond
an iceberg, over two miles away. With the aid of the glass
we could see it was moving in our direction, and we thought
it was Annowkah coming back from the other bay. Kyo, who
was watching constantly, all at once became very much excited,
declaring it was not an Innuit, and he could not tell what
it was. Then, suddenly throwing down the glass, his eyes
almost starting from his head, he exclaimed, " Nahnook, nahnook, boo mut toy-hoy, car, car, toy-hoy " (a bear! a bear!—
the rifle, quick, hurry, hurry, quick). Matt and I rushed into
the house for our rifles and ammunition, but by the time we
came out the object was behind the berg, lost to view. It soon
reappeared, however, and we then saw that it was a dog. Kyo,

who had been watching it closely, immediately recognized it as one of Mr. Peary's pack, and said that it was in a starving condition. The poor animal was hardly able to get along, and had evidently had nothing to eat for a week or ten days. He is very weak, especially in his hind legs, and he has a cut from his left eye down to his mouth. The dog is the one which we had designated the " devil dog," and was in charge of the supporting-party. Can it be that the supporting-party has met with mishap, or are they returning by way of Smith Sound? The incident brings up unpleasant forebodings, but I am utterly powerless in my position.

Thursday, June 2. Three more days of increasing suspense, and still no news. It is now twenty-seven days since Gibson left us to rejoin the party, and at that time Mr. Peary wrote, " We go over the ice-cap to-night," and he thought that the supporting-party would be back in ten days, or at most in two weeks. Spring is now rapidly coming to us, and the mercury, in the sun, has risen well into the seventies.

Friday, June 3. My nightmare is over; the boys have returned, and they bring good news of my husband. I cannot describe how I felt when the doctor, on shaking hands with me, told me he had left Mr. Peary and Astrup both in good health and spirits, and doing good traveling. Both boys look exceedingly well, although their faces, and noses particularly, are much burned and blistered by the sun and wind, and Gibson complains of his eyes. I got them something hot to drink, made them chocolate, and then retired to my

room to read my letter.   Gibson weighs 173¼ pounds net,
against 176¼ when he left; the doctor weighs 153 pounds
net, as against 146¼.

Saturday, June 11. The past week has been almost entirely without incident. Dr. Cook has assumed command of
our establishment, and I am therefore free of responsibility
beyond that of taking care of myself. My thoughts wander
constantly to the members of the inland ice-party, and I often

A Corner of my Room.

wonder if they will return in time for
us to go south still this summer. The
doctor and Gibson do not expect them
before the 1st of September, while our
Eskimo friends cheerfully assure us that
they will never return. My instinct re-
volts against this judgment, but it makes
an impression upon me, nevertheless.
To-day I walked over to the Quarter-
Mile Valley, and sat by the stream which
there rushes down from the cliffs and tumbles over the icy
hummocks, cutting its way through the snow that fills its bed
and over the ice-foot into the bay. The little snow-buntings
were chirping and flitting about me, and great patches of
purple flowers, the first of which I observed just one week
ago, were to be seen wherever the snow had melted sufficiently for them to peep through; these were the earliest
flowers of the season. I sat here and indulged in a fit of
homesickness. Never in my life have I felt so utterly alone

and forsaken, with no possible chance of knowing how and where my dear ones are. It surely must end some time.

Sunday, June 12. The snow is disappearing rapidly, and just as soon as a patch of ground is laid bare it is covered with flowers, usually the purple ones, although I have seen a few tiny white and yellow ones as well. The west wall of our entrance is covered with green shoots. The doctor and Gibson are preparing for a ten days' hunting-trip up the bay, and they have made up the following list of provisions and accessories: 140 crackers (seven per man per day), 10 pounds sugar, 4 pounds meal, 8 pounds hominy, 5 cans milk, 1 three-pound can of tongue, 2 cans corned beef, 3 cans tomatoes, 3 cans corn, 2 cans soup, 4 cakes pea-soup, 4 pounds bacon, 1 package cornstarch, 1 can Mosqueros food, flavoring extract, salt, 4 pounds coffee, ½ pound of tea, 15 pounds dog-meat for two dogs, 2 cans alcohol, 2 alcohol-stoves, 2 boxes wind-matches and 1 box blueheads, 1 box of cartridges, and a number of shells. They expect to leave this evening. The condition of Matt's frozen heel has been steadily growing worse, and, poor fellow! he is beginning to suffer acutely. He is threatened with a chronic running sore.

There is only one thing now left to me which gives me any pleasure, and that is to go to the little brook in the Quarter-Mile Valley and listen to its music while I give my thoughts full play. I close my eyes, and once more I am in our little tent, listening to this same music, mingled with the sound of the "Kite's" whistle and the splash of the white whales as they frisked back and forth in the water close to the shore. This was when we first landed, and before the house was ready for us.

Wednesday, June 15. The last of winter is leaving us. The water is rushing and gurgling on all sides, and the brown cliffs back of the house, as well as the red cliffs to the right, are almost entirely bared of the snowy mantle which has so long covered them. Eider-ducks are passing us daily, and in their wake come other birds from the balmy south.

My routine tramps have been largely interfered with by the character of the walking, which has become very bad, snow, slush, and water alternating in layers. Into this one plunges thigh-deep without warning, and it requires considerable maneuvering to extricate one's self without becoming saturated with ice-cold water. The tide comes in beyond the ice-foot, and Verhoeff almost swims to the tide-gage, which is now five inches higher out of the ice. I have been for some time past taking my watch regularly with the boys, and naturally it interferes somewhat with the fulness of my night's rest. At present the night is divided into three watches, of which I take the first, Verhoeff the second, and Matt the morning watch.

Wednesday, June 22. Another week has passed, and by this much my husband is nearer to his return. Our routine continues unchanged, except in unimportant details, and the monotony of our life, together with certain vexations which necessarily arise, makes me at times cross and despondent. Our Eskimos have been taking advantage of the open leads

and the return of animals to go out on various hunting-expeditions, and they report more or less success with walrus, white
whale, and narwhal.   I am longing for venison, as we have
been largely reduced to a seal diet, and seal is all but nause-
ating to me.   Deer seem to be very difficult to get at just at
present, and Dr. Cook, who returned early Sunday morning
from his hunt at the head of the bay, brought none with him
—indeed, no meat of any kind.

The first rain of the season took place last Thursday night,
and it has been raining again lightly this evening.   Yesterday
I took a walk along the base of the trap-dyke.   The snow has
disappeared from the plateau, and the air is fragrant with the
spring flowers and mosses, which fairly cover the ground.
Numberless snow-birds are flitting about, chirping to each
other, and the rushing of the brooklets is heard constantly.
All the flowers have returned and all the birds are here again,
and they will stay with us until the middle of September, when
I hope that we, too, shall return south.   Altogether the scene
reminded me of the time when Mr. Peary and I came up here
last fall, and I gathered flowers while he pressed them.

Tuesday, June 28.   What a horrible day it has been!   The
wind blows so hard that it is almost impossible for me to
stand up against it.   The rain dashes against the window until
it seems as though it would break it in.   At times the rain
changes to snow, while on the cliffs it has been snowing constantly.   They are as white as they have been any time this
winter.   Icebergs have been groaning and toppling over all

11

day, and in the fury of the storm, just after midnight, the tide-gage fell over. My constant thought is of the advance party. God help them if they are caught in such a storm on ice that is not suitable for building igloos. As the days wear on I feel as if the chances were almost even as to whether I shall ever see my husband again. I can do nothing, not even keep still. Perhaps it is a good thing that I am obliged to do the work about the house.

Our boys have been improving the time by gathering up collections of various kinds, and the doctor has been especially busy trading for any and every thing in the way of native clothing, implements, and toys, for all of which he gives pieces of boards, barrel-staves, boxes, and other odds and ends in the lumber line, all worthless to us, but invaluable to the poor Eskimos. Wood is to them their most precious article, for without it they could neither have boats nor sledges, nor would they be able to fashion those perfect instruments of the chase, the harpoon and spear, which they handle with unsurpassed dexterity. Yet wood is also their scarcest article, and is obtained only from wreckage or through occasional barter with whalers passing near Cape York. A cargo of lumber would procure anything from the natives— indeed, almost their entire possessions.

Friday, July 1. To-day we narrowly escaped a bad accident. The doctor accidentally discharged a gun in the big room, where Gibson, Verhoeff, and Tooky were sitting. Fortunately no one was hurt, the charge going through the roof, making quite a hole, and badly frightening Matt, who was lying there. Matt's foot is improving somewhat, and probably in a few days his condition will be such that he will be able to get about. This prospect is gratifying to me, as I have determined to go to the head of the bay in about three weeks, there to await Mr. Peary's return, and I wish to have Matt for my companion.

Monday, July 4. This evening I was treated to a native vegetable dish. Returning from a walk to Cape Cleveland, I met Mané and her children coming to meet me. She told me they eat the little purple flowers which bloom so abundantly almost everywhere in this vicinity, and asked me to try them. I found that they were quite as sweet as our clover blossoms, and they have, besides, a very aromatic flavor. Mané had brought two of our tin mess-pans with her, and we filled them with blossoms and sour-grass. On reaching Redcliffe Mané mixed the flowers and sour-grass, then, pouring a little water on them, put them on the stove. I suggested that she wash them so as to remove at least some of the sand, at which she laughed, saying that sand was good for the stomach; nevertheless, she made a show of washing them, and then let them boil for about fifteen minutes. The flavor was a peculiarly pleasant one, but I thought it a little sour, and added some sugar, which gave it something of the taste of rhubarb-plant stewed, only more aromatic.

This concoction is the only vegetable dish that these people ever have, and this is only eaten by the women and children, not by the men. On the other hand, the men eat the eggs of the different birds, but will not allow the women to touch them. It was amusing to see both Mané and M'gipsu eat cake containing eggs, begging us not to tell their husbands, and consoling themselves with the reflection that eggs did not form the chief part of the cake.

Wednesday, July 6. Another sunshiny day. Yesterday morning two Eskimo boys came in, and reported that a whole troop of natives were at Ittiblu on their way over from Netchiolumy. They are compelled to go up the gulf this far in order to cross on the ice above the open water.

The open water has now nearly reached Redcliffe, and is full of birds. About five o'clock this morning fourteen natives arrived, among whom are Mekhtoshay (the one-eyed man) and his wife and boy, and Ingyahpahdu and his six children. The one-eyed man brought his tent with him, a very small one, but the others are camping with their neighbors—a privilege which is generally permitted in traveling. We have taken advantage of these numerous arrivals to continue our series of ethnological photographs, and the doctor has been kept busy posing, grouping, etc. Our settlement now numbers thirty-four natives, men, women, and children.

Gibson has started off on a ten days' collecting-tour to the head of the bay. He will leave the tent in Tooktoo Valley for me, and I shall go as soon as he returns, taking provisions enough to last till August 6th. If Mr. Peary has not returned by that time then I shall come back to the house and get everything ready for our homeward journey in the early autumn.

Thursday, July 7. I determined to take advantage of the fine weather we are having and get rid of some washing to-day. I also put Noyah, Mané's little one, in the tub and gave her a good scrubbing. She actually looked quite cute, and after getting over her surprise at being plunged into the water, enjoyed it, laughing and splashing. It seems odd to see the children so backward. This child, who is already two years old, has just begun to stand alone, and in all other respects she is like a child at home of ten months or a year. M'gipsu's baby is a year old, but in size and mental development compares with a five-months-old white baby. To-night we finished taking the photographs and measurements of the Eskimos.

Sunday, July 10. The day has been bright, warm, and sunny. At eight o'clock this morning the thermometer in the sun registered 92°, and still it would be called a cool, pleasant day at home. The doctor tore down the shed back of my room in order to give the sun a chance to melt the ice and dry the things under it.

Ikwa killed an "oogzook" this morning while out in his kayak. It took three men all day to bring in the skin and part of the carcass. Ikwa says he has to divide the skin among all the men in the settlement, even Kyoshu the cripple coming in for a share. It is the rule that every animal killed, larger than a seal, must be divided among all the men in the community, regardless of their share in the securing of it.

Monday, July 11. When I awoke this morning I heard Matt and the doctor talking very earnestly, but could not

11*

hear what they were saying; from their tone I judged it was something serious. Finally I called to the doctor and asked him what the trouble was. He told me that Matt had overheard Kyo and Kulutingwah planning to make away with one of us. I could not help laughing at this recital, which provoked the doctor a little; we had laughed at similar stories related by Arctic explorers, and had agreed that these natives were not at all inclined to be warlike or vindictive. I tried to reason with the boys. In the first place, if the natives had any such design, would they not have kept the three men here who left for Karnah yesterday?. Secondly, would they be likely to come over to our house and discuss their plans? And thirdly, do any of us know enough of their language to understand a conversation in which the participants are not even to be seen? The whole thing seemed very amusing to me, but both boys were evidently frightened, and wanted to be armed and ready for any emergency; consequently, I gave the doctor Mr. Peary's pistol to carry and Matt my large one, and they have worn them all day. Matt imagined he knew the cause of the whole thing, namely, Kyo was mad because I had stopped his coffee and bread in the morning; he had blamed Matt for it, and so Matt felt certain he was to be the victim. The fact is, however, that Kyo got his coffee as usual this morning. I had intended to stop it, but as Mané was sick and did not care for her share, there was enough to go round. The doctor, more than any one else, has reason to fear Kyo, as Kyo makes no secret of his dislike for him.

One year ago to-night was the most miserable night I had ever spent. Mr. Peary had broken his leg, and for a few hours I did not know whether he would ever be able to use it again; to-night I do not even know that he is alive. I feel very certain, however, that a month will solve this question for me, and so am determined not to worry any more.

CHAPTER XV

MY CAMPING EXPERIENCE IN TOOKTOO VALLEY

Conclusion of the Murder Scare — A Fifteen-mile Walk along the Arctic Shore — Matt my Sole Companion — An Arctic Paradise — A Tramp with an Unpleasant Ending — Twenty-four Hours with Nothing to eat — In the Shadow of the Ice-cliffs — Fording a Glacial River — Safe in Camp again.

Tuesday, July 12. Gibson arrived this morning, minus his sledge and his entire load, having been obliged to abandon them on account of hard traveling. He advises me to go to the head of the bay without delay, as the ice is even now in a bad condition, and each day makes it worse. Ikwa was on the point of starting with a sledge of provisions and bedding, and I decided at once that Matt should accompany him. I shall follow later along the shore. At one P. M. Matt and Ikwa started, with five dogs, one native sledge, and one toboggan. I fully intended to leave after supper, but I found so many things to do that I was too tired to think of walking fifteen miles, and determined to wait until to-morrow. I gave my room a thorough cleaning, and put down my new carpet, washed and did up my bed-curtains, and made things as bright and clean as possible. I hope the little den will look somewhat homelike to Mr. Peary when he comes back.

A FRIENDLY "TUPIC" AND ITS INHABITANTS.
(Looking out of McCormick Bay.)

I am afraid this lovely weather will not last much longer; but even if it rains I believe I can be as comfortable in the tent as here at Redcliffe.

Kyo came in to-night and had a long talk with the doctor about the doctor's threatening to shoot the huskies. He is very much frightened at the doctor's carrying the revolver. What added to his fright was that we opened the side window this afternoon, Kyo immediately concluding that we intended to fire on the natives from it. I am more than ever convinced that there was nothing in Matt's "overheard conversation," and it is certain that all the Eskimos are badly frightened at the display of firearms. Kyo said the doctor might shoot the others, but the bullets would not hurt him; that the "kokoyah" (evil spirit) was kind to him, and he would never die. But if the white man killed the Innuits the kokoyah would, at Kyo's command, "shad-a-go" (destroy) their vessel, and they would all die. Finally peace was declared, and Kyo brought over his sealskin float, for which he wanted wood to make the ring of his kayak. I am sorry for this episode, which has brought about an unpleasantness with the natives.

Wednesday, July 13. At 2.30 this afternoon, in company with Dr. Cook, I left Redcliffe on my fifteen-mile walk to the head of the bay, which we reached at eight o'clock. Matt and Ikwa, who had preceded us, had a terrible time in getting through. Half the time they were in water above their waists, and occasionally they were obliged to float themselves over on

Ikwa's sealskin float. It was all that Matt could do to persuade Ikwa to continue. It began to rain about ten P. M., and has rained lightly ever since. I fear the doctor did not have a pleasant walk back.

Thursday, July 14. I made a short scout after duck, but saw only a few eiders far out on the ice. How sweet the air is, and how restful the rushing of the streams as they make their way to the shore! I feel the need of rest and quiet, and it is very peaceful here. When the weather clears I shall enjoy the rambles over the soft green moss, I know.

Friday, July 15. This morning the sun was shining brightly, and had it not been for the mosquitos the day would have

been thoroughly enjoyable. Matt and I started about nine A. M. to take a look at the country beyond Boat Camp, but I find it will be impossible to

A Garden Spot.— Greenland Moss and Poppies.

cross the glacial river, and yet I must get to Tooktoo Camp before long. After lunch I took my shot-gun and started out in the direction of the hanging glacier, where there are a number of ponds. In one of these I saw two long-tailed ducks, but I could only secure one. The breast gives us one meal, and the rest of the bird stew for another. After supper we took a walk over the hills toward the glacier. The evening was fine, the air sweet, the grass and moss soft, and studded with thousands of flowers. In every direction can be heard either the rushing and roaring of a glacier river, or the rippling and swishing of some tiny stream. The snow-buntings and sandpipers are hopping about and chirping merrily, and the great golden ball is moving slowly along the heavens. The inland ice seems to wear a continual smile, so bright does its surface appear. Does it wish to assure me that all is well with the ones who are traveling on its bosom, or is it only mocking me? I will try to think the former.

Sunday, July 17. A dull, foggy day. The mosquitos are so thick that it is all but impossible to venture out.

Wednesday, July 20. Yesterday at noon the sun was shining brightly, and there was a light southeast wind, enough to keep the mosquitos quiet, so I decided to start for the cache back of Tooktoo Camp, in which I wished to deposit a note and some canned goods. I knew it would be a long tramp around the intervening lake, but I would be amply repaid if my husband were to return while I was still here, and find the note, assuring him of a welcome a few miles beyond. When we reached the mouth of the glacial stream which discharges into the head of the bay, it was low tide, and we

made an effort to ford it, thinking thereby to save a walk of
five miles. Matt stepped in and I followed. The water felt
intensely cold; it was above my kamik-tops, but not above
my knees, and we went on. When we came to a rock about
one fourth of the way over I was compelled to climb on it and
beat my feet and legs; I could not control them any longer.
Then we again plunged into the icy water, which now reached
above my knees. It took us fifteen minutes to cross, and the
temperature of the water was certainly not over 35°, for large
and small pieces of ice were floating about us. The current
was in places very strong, and had it not been for the boat-
hook I had taken with me, on which to hoist a flag over the
cache, I should have been swept off my feet many times.
Once across, and our wet stockings changed for dry ones, I
did not regret having come. We found the cache after some
little trouble, and I deposited the note, also a can of milk, a
can of fruit, some biscuit, and a small flask of brandy, and
then put up the flag.

We retraced our steps past old Tooktoo Camp to the mouth
of the river. Here we found that the tide had already risen
a foot, and we continued our walk along the river-bank toward
the head of the lake. On reaching it we found that it com-
municated with a second lake by a deep, roaring torrent, which,
although narrower than the river below, was still too wide and
deep to be crossed; so on we went till we reached the end of
the second lake, and here it seemed as if we might walk around
it by climbing along the lower edge of two glaciers, although

we were by no means sure that a raging stream did not sweep down on the other side. Great rocks were continually rolling from the top of the glaciers, and I did not think it safe to venture. The scene was an impressive one. Black cliffs raise their heads over four great white glaciers, smooth as marble, and at their feet roars a furious torrent, till it merges into a broad lake, which looks as calm and unruffled as if this stream were only a drop in its depths. On each side of this stretch of water the valley is carpeted with soft green moss and yellow poppies, and fairly alive with the chirping and flitting of birds. We tarried here quite a while. I could not make up my mind to leave so beautiful a scene; besides, the only thing left for us to do now was to wait for low tide, which would be about one A. M., and then ford the river where we had crossed it in the morning. It was 8.45 P. M. when we again reached the mouth of the stream. The tide was high, but falling. Had we had something to eat we should not have minded the waiting. We kept moving in order to keep warm, until we thought that the tide had reached its ebb. As we neared the shore we could see no familiar line of rocks which indicated low tide, and on closer examination we were horrified to find a "high low tide." Still we felt we must attempt to cross, and Matt started in, while I followed at his heels. The first step was over our knees, the next came mid-thigh on Matt, and then I backed out, for I knew that we were not near the deepest part yet; besides, the current was so strong that I could hardly keep

my footing. We tried lower down, but with the same result.
Even had we made up our minds to bear the cold water, we
could not possibly have stood up against the current. We
then determined to try it in the lake, but were baffled there as
well. By this time we were pretty well drenched, almost to
our waists, and yet the only thing for us to do was to wait for
the noon low tide of the morrow. We sat down on a rock,
took off our stockings and kamiks, and wrung the water out
as best we could, then put them on again. I knew it would
never do for us to sleep, or even sit still in our wet clothes,
for there is always a cool breeze blowing, and the night temperatures average about 40°; yet the prospect of twelve
hours more of tramping, when we had already tramped
twelve and a half hours, with nothing to eat—we had only
had coffee and a cracker before starting—and a cold fog set-
tling down upon us, was anything but encouraging. I suggested that we go to the cache, where we had left the brandy
and milk for the inland ice-party, and mix a drink of some of
it, and then begin the climb to Nunatak Cache. This we did.
I had my old enemy, the sick headache, brought on by lack
of food and the excitement, and consequently every step was
agony, yet I knew I must keep on. Thoughts came crowding in upon me of my husband and my mother. We walked
and walked until almost ready to drop with hunger, fatigue,
and lack of sleep; then, as we climbed above the fog into the
warm sunshine, we would sit down a few minutes, wrapping
our heads in our handkerchiefs to keep off the mosquitos,

which swarmed about us. As soon as one of us saw the other dozing we pushed on again. In this way we climbed through the ravine and in sight of Nunatak Cache, but it was impossible for me to go farther; my limbs trembled under me, and refused to act at my bidding. We returned to the river. At 11.30 this morning the welcome line of rocks indicating low tide made its appearance, and, to our great relief, we found that we were able to cross the stream. Two more thankful creatures never were than we when we found ourselves on dry land on our side of the " kook " (river) again. We were perfectly numb with cold from mid-thigh down, and so ran and pounded our feet and limbs for the three miles that intervened between the river and the tent, which we reached in an hour. Thus far we feel no ill results from our icy adventure.

Saturday, July 23. The bay, which has been perfectly clear of ice, except for a few small bergs near the glacier, is filled again, as a result of the tide-wind. The white whales, which have been sporting about for a number of days, are shut out from their playground. I tramped about nearly all day, but did not get near any game. I never weary of Tooktoo Valley. To me it is a beautiful spot, with its river and lakes, its glaciers and mountains, its carpet of soft green moss, its wealth of flowers, and its busy birds and insects. I have not heard from Redcliffe since I left there, over a week ago; no information of any kind has come to me.

CHAPTER XVI

"OOMIAKSOAK TIGALAY!"—THE SHIP HAS COME!

An Eskimo Messenger — "Oomiaksoak Tigalay" (the Ship has come) — Letters from Home — A Visit from Professor Heilprin — Distressing Possibilities — The "Kite" leaves for Smith Sound — Return of the "Kite" — Domestic Disturbances among the Natives — An Eskimo Woman and Girl disappear.

Sunday, July 24. At five o'clock this morning, before I was really awake, I heard a sharp, shrill whistle, different from the notes of the birds that usually awake me, and before I could quite satisfy myself that it was not a bird I heard it again, close to the tent, and also a footstep. "Kiny-ah-una" (who is there), I called. "Awangah, oomiaksoak tigalay" (me, the ship has come), was the answer. "Angwo" (not so), I replied. "Shagloo nahme awangah" (me not lie), he said, and with this a shaggy, black head was thrust into the tent, and a bundle of mail tossed to me. The next few hours are a blank to me, for I was devouring my mother's letter, which took the shape of a journal that she had kept for me. A few words from Professor Heilprin tell me that he is at Redcliffe with a party and the old "Kite," but he does not say who are in the party. Now if Mr. Peary only gets back safe I shall indeed be happy. All those dear to me have been spared,

ROBERT E. PEARY, U. S. N.

while there has been a great deal of sickness and death everywhere.

Monday, July 25. This morning the sun came out bright,
and he has shone all day. After looking in vain for the
inland ice-party, and also for a party from the "Kite," until
two P. M., I retired to the tent to escape the mosquitos. I told
Matt he might go down to Redcliffe and see the "Kite" party
if he chose, but he said he did not care for the walk, and would
take the gun and go for a stroll. At 3.30, feeling hungry, I
went out to see if I could see anything of him, in order to
know whether I should cook for one or for two. Away off
near the foot of the cliffs I saw a lone figure, which did not
look like Matt, slowly making its way in the direction of the
tent. I soon made out Professor Heilprin. He had walked
fifteen miles to pay me a visit, and we chatted for hours. It
did seem so good to talk with some one again who had been
in touch with civilization. I feel as though I had been in
another world. Both mother and brother urge me to come
home, even if Mr. Peary has not returned from the inland ice
by the time the "Kite" is obliged to set sail again for the
sunny south, and the professor says his orders are to "bring
Mrs. Peary back under any circumstances." While I do not
think there is the slightest doubt that my husband will be
here before the latter part of August, and while I fully believe
that if he is not here then he will never come, yet I could
never leave while there was the faintest chance of his being
alive. I told the professor just how I felt about the matter,

12

and he said, "Well, we will see when the time comes." My brother Emil writes that I should have "some consideration for my friends and relatives." And what of my husband? He says further, "What good can you do Bert on the coast while he is on the ice?" Does he suppose that if Mr. Peary is alive he will stay on the ice the whole year round? And when he returns and finds he is too late for the "Kite," will that not be disappointment enough, without finding that I, too, have deserted him? I know just how my dear ones at home feel, and I know, too, that they cannot long for me any more than I long for them. It will go hard to remain — harder for me than for them, for they will know that I am well and comfortable; and besides, they have friends and acquaintances, and intelligent and interesting employments and amusements with which to occupy their minds and time, while I have only a few white men and some uncivilized people, together with three months of darkness, to make my life pleasant. Not a very enviable existence, I am sure. As for cold, hardship, and hunger, that is nonsense. Of course, if I feel so inclined, I can go out and sit on an iceberg until I freeze to it, and let the wind and snow beat upon me, even starve myself; but my tastes do not run in that direction.

Tuesday, July 26. The "Kite" leaves to-day for Littleton Island, to be gone three or four days. When the professor left, at 2.30 A. M., Matt had not yet returned; I think he must have gone to the "Kite."

Wednesday, July 27. Yesterday and to-day were bright,

warm days, although the wind blew quite strong most of the time. Matt returned from the "Kite" yesterday morning, bringing with him a loaf of nice bread, a veal cutlet, and a flask of brandy sent by the steward of the "Kite." Dr. Cook, with four Eskimos, came up in the "Mary Peary" this morning, bringing the rest of the mail matter with him. He also brought me more supplies, but at the same time urges me to return to Redcliffe with him.

Saturday, July 30. Once more back at Redcliffe. After considering the matter, I decided that Mr. Peary would wish me to look after things at our home, and although it was a great disappointment for me to leave before the return of the ice-party, I was forced to do it. There has been considerable excitement in our Eskimo settlement. Ikwa has beat Mané so badly that she cannot come out of her tent; her head is cut and bruised, and one eye is completely closed. We know of no reason for this peculiar conduct. Kyo has gone to Igloo-dahominy in his kayak, the first time during our visit that an Eskimo has ventured across the bay in a kayak. While he was out on a seal-hunt early this morning, Klayuh, his wife, and Tooky, her daughter, ran away. Kyo, it is said, had thrust a knife in Klayuh's leg several times, and he has threatened to kill Tooky. He is now searching for the fugitives, but the whole settlement has conspired to throw him off the track. He has already been up to the head of the bay, and down as far as Cape Cleveland.

The "Kite" returned at nine o'clock yesterday evening,

having penetrated into Smith Sound to a position opposite
Force Bay, where it was stopped by the unbroken pack.
Professor Heilprin came ashore immediately after, and intro-
duced to me some of his companions.   Dr. Cook, who had

The " Kite " in McCormick Bay.

made a vain attempt to reach Ittiblu, returned at ten P. M.
this evening; he found the gulf impassable owing to the large
quantities of loose ice which had been detached from the glaciers, and literally choked the basin.

Thursday, August 4.   I have lived through five days more
of intense suspense.   The Eskimos console me by talking of

Mr. Peary as "sinnypoh" (dead); one of them yesterday told me that he had dreamt that only one "kabloona" (white man) would return from the ice. To offset these dark forebodings, and keep my spirits from sinking too low, I repeat a paragraph in Mr. Peary's letter, which says: "I have no doubt I shall be with you about August 1st, but if there should be a little delay, it will be *delay only*, and not danger. I have a hundred days' provisions."

The weather continues exceptionally fine, clear, bright, and warm. Professor Heilprin, having determined to move his party to the head of the bay, preparatory to a search on the inland ice, the "Kite" heaved anchor at nine this morning, and is now lying opposite the point which I only recently deserted. By the professor's kind invitation I joined the "Kite" party, and Matt, who has been my steady guardian since Mr. Peary's departure, accompanies me.

Friday, August 5. The entire relief-party left to-day for Nunatak Cache, their object being to plant stakes seven miles apart as guide-posts on the inland ice. I remained on board the "Kite" all day, and shall retire early, if not to sleep, to rest.

CHAPTER XVII

RETURN OF THE EXPLORERS

End of my Weary Waiting — Mr. Peary returns " on Time " — Experiences of
the Inland Ice-party — The Great Greenland Ice-cap — The " Kite " Aground
— Landing through the Surf — Back at Redcliffe — The Natives regard the
Commander and Astrup as Supernatural Beings.

Saturday, August 6. From a half-sleep I was roused early
this morning by the plash of oars and loud talking, and before
I had fully grasped the idea that the professor's party had
returned, some one jumped over the rail on the deck just
over my head, and a familiar footstep made its way hurriedly
toward the companionway. I knew it was Mr. Peary, but
was unable to move or make a sound. He came rushing
down the stairs and rattled at my door, calling to me to open
it; but I seemed to be paralyzed, and he forced it open and
stood before me, well and hearty, safe at last.

Monday, August 8. Back at Redcliffe again, but how
different everything seems! Not only is our whole party
once more reunited, but there is the little " Kite " out in the
bay, ready to take us south at any time.

I have been afraid to go to sleep since Mr. Peary's return,
for fear I might wake up and find it all a dream; besides, we

had so much to tell each other that there was no time or incli-
nation for sleep. Mr. Peary recounted to me the events of his
journey; how after he sent Mr. Gibson and Dr. Cook back
to Redcliffe from the Humboldt Glacier, May 24th, he and
Astrup marched on day after day, with their magnificent team
of Eskimo dogs, which Astrup learned to handle as well as a
native driver.

They encountered storms which kept them buried in the
snow for days at a time, but their worst enemies were the
snow-arched crevasses which they
met just before reaching the latitude of Sherard Osborne Fjord.
These arches were so treacherous
that more than once they were on
them before they were aware of
it, and old Lion came very near
ending his journey
by breaking through
one of them and
being precipitated
the full length of
his trace into the
yawning chasm.
Fortunately the
trace was strong
enough to hold his
weight, and he was

The First Musk-ox.

pulled up none the worse for his tumble.   The loss of a single
animal would have been a calamity to the party.

On July 1st Mr. Peary saw the loom of land ahead, and
thinking it only one of the west-coast mountains, changed his
course to northeast, and then to east, hoping to be able to
round it; but after a few days' further travel he saw the land

Cairn on Navy Cliff.
Lat. 81° 37'.

still ahead, and then found
that it was the northern
boundary of Greenland.
He now decided to leave
his sledges and supplies at
the edge of a moraine, and,
with a few days' rations, start overland toward the coast.   They had not
gone far when they came across unmistakable signs of musk-oxen, and
then upon the animals themselves,
grazing in a little valley. A few shots
from Mr. Peary's rifle brought down
two of them.   Then a little baby musk-
ox came peering around a great boul-
der to learn the cause of all the noise
and commotion.   This was captured
alive, but the poor little thing did not
survive its mother very long.   Mr.
Peary camped in this lovely valley, and there feasted his dogs
on fresh meat.

ACROSS THE SNOW DESERT.—FOLLOWING THE GUIDON.

These noble brutes, accustomed all their lives to raw, bloody meat, had been living on dry pemmican for the past two months, working day after day as they had never worked continuously before. No wonder they strained at their traces, plunging and tugging to get loose and help themselves. As quickly as one of the musk-oxen was skinned the body was tossed within their reach, and they pounced upon it with a greediness which plainly showed how much they longed for the juicy meat. The explorers themselves also enjoyed the fresh meat for a change, but they were glad to get back to pemmican again after a few days.

After the dogs had been fed and rested, the march across the boulder-strewn country toward the coast was resumed. It ended July 4th, when the party came out on a bluff on the east coast, some 3800 feet high, which overlooked the great unknown Arctic Ocean. Here a couple of days were spent in making observations for latitude and longitude, in taking photographs of the surrounding country, and in building a cairn in which to deposit the record of their journey, and then the return march was begun. McCormick Bay was reached on August 6th, after an absence of ninety-three days, during which time Mr. Peary says neither he nor Astrup had an ache or a pain.

Late yesterday afternoon a brisk wind blew up that made the surf fly and prevented any of us from going ashore. As Professor Heilprin was anxious to examine some of the great glaciers, it was decided that the " Kite " remain at her present anchorage until after he had made his examinations the next day. This morning, however, the wind was still blowing, and although an attempt was made to land a boat, it had to be abandoned; Captain Pike, too, was desirous to get the "Kite" down the bay before she was blown on the rocks. Indeed, this was necessary, as the vessel had already had her nose stuck in the mud-bank, and it had seemed for a time that she was in a precarious position.

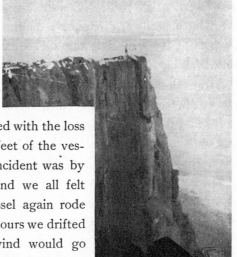

Fortunately we escaped with the loss of only about eleven feet of the vessel's "shoe." The incident was by no means pleasing, and we all felt relieved when the vessel again rode a straight keel. For hours we drifted about, hoping the wind would go down, but finally we headed down the bay. It was impossible to swing the vessel inshore opposite Red-

Looking down over the Arctic Ocean.

cliffe, and we were obliged to pass our home and continue to Cape Cleveland. Here again we could find no sheltered nook where it would be safe to land a boat, and we sailed back and forth until late in the afternoon, when the captain thought that we might land in the lee of the great cliffs just east of Cape Cleveland. The boat was put in charge of the second mate, who, with the three strongest sailors, pulled Mr. Peary, Astrup, and myself to the shore, a distance of perhaps half a mile. We got along well in spite of the great billows until we reached the shore, where, before we could make a landing, two waves in rapid succession broke over our boat, almost filling it with water, and nearly swamping us. I was completely drenched.

Just before reaching Cripple Point we were met by Dr. Cook, Verhoeff, and Gibson, anxious to greet the inland ice-party, of whose return they had been apprised by Matt. It was very curious to watch the expressions on the faces of the natives, who stood in groups about Redcliffe House staring at Mr. Peary and Astrup as they approached. When they were spoken to they answered in low, frightened tones, and they could not be induced to come forward and shake hands, or in any way come in contact with the two, until they were convinced that they were really human beings, and not great spirits come down from the ice-cap. Then they were very anxious to know if Mr. Peary had seen the spirits of the departed Eskimos, what they lived on, how they looked, and all about them. They were very much surprised not only to see the dogs return alive, but to see them in much better condition than when they left, as they had repeatedly said the Americans did not know how to feed the Eskimo dog, and he would soon starve under their treatment. Now they have perfect confidence in Mr. Peary, and say they would go any-where with him, even on the ice-cap, because they do not believe he would let the evil spirit harm them.

Mr. Peary has decided to start on a trip up Inglefield Gulf to-morrow. His purpose is to verify some of the observations made by us on our April sledge trip, to take photographs of the landscape in its summer dress, and to secure ethnological specimens at Karnah and Nunatochsoah that were promised us by the natives of those places. We ex-pect to return within a week, and then everything will be put on board the good ship "Kite,"

Astrup's Musk-lamb.

and we shall bid adieu to our Arctic home and the dear old huskies, who, even if they are not particularly clean, have been our faithful friends, and will, I am sure, never forget us.

IN MUSK-OX LAND BEYOND THE ICE-CAP.

CHAPTER XVIII

BOAT JOURNEY INTO INGLEFIELD GULF

The Sculptured Cliffs of Karnah — Luxuriant Vegetation — Stormy Weather — Anniversary Camp — My Kahlillowah — Crossing the Gulf in a Tempest — The Shelter of Academy Bay — Fury of the Arctic Winds — An Iceberg Breakwater — We reach Karnah again — Rounding Cape Cleveland — Fighting for Life and Shelter — Safe at Redcliffe.

The weather was not very encouraging as we started from Redcliffe House on Tuesday, August 9, the strong wind of the two previous days having brought up heavy storm-clouds, which now hid the sun and hung threateningly overhead. It was just about noon when we left the beach at Redcliffe, the light " Mary Peary " shooting rapidly along with the strokes of the six Eskimo boatmen, and in a short time we had rounded Cape Cleveland and started eastward up the gulf. The scene before us was very different from what it had been ten months previously, when we made our first attempt. There were then numerous pans and streams of ice, with the new ice rapidly cementing them together; the land itself was covered with snow, and the ice-foot had already commenced to form on the beach. Now there was only an occasional fragment of ice, though the great bergs were numerous. The mountains of the shore were rich with the warm hues of summer. Late in the afternoon a favoring wind came up from the west, and with foresail hoisted we moved merrily along before it. Relieved thus from their labors, our crew lounged contentedly upon the seats, and fell into a conversational mood. Mr. Peary learned from them that many years ago Mekhtoshay had shot an " amarok," or wolf, at Netchiolumy, and that Panikpah had killed one at Nerki; Koomenahpik and Mekhtoshay, who are brothers, also related that years ago they had both seen " oomingmuk" (musk-oxen), " awahne, awahne, Etah " (far beyond Etah).

At half-past six in the evening we reached Karnah, a small Eskimo settlement on the north shore of the sound, some twenty miles from Cape Cleveland. Here the low, flat shore ends, and a line of towering gray cliffs begins. We pitched our tent on a level bit of grass among the stones, and after our evening meal was completed we crossed the noisy glacial stream flowing near the village, climbed the hill just west of it, and then followed the shore westward till we came to the stone igloos of Karnah the deserted. Four houses form this village, which lies in the center of a beautiful grassy meadow, stretching back from the shore to the foot of the brown mountains. The luxuriance of the grass here was wonderful. All across the meadow we waded through it, literally knee-deep, and in one or two places immediately about the igloos it was so rank that as I stooped to gather some sprays for pressing I was almost hidden. Returning to our tent, we were soon lulled to sleep by the boisterous music of the glacial stream. During the night it snowed lightly, and when we awoke the ground was covered with a white mantle, which, however, soon disappeared.

Leaving Karnah on the morning of the 10th, for three or four hours we threaded our way through bergs and great cakes of blue ice, past the giant cliffs of Karnah, with their great bastions, towers, chimneys, and statues, carved by the Arctic storms from the gray sandstone, the breeding-places of black guillemots, burgomaster gulls, and white falcons. As we passed along our Eskimo boatmen pointed out to us the striking figures, all of heroic size, looming against the sky far up the cliffs, and told us that such and such a one was a woman, and such another a man, and that the cliffs themselves were known as "innuchen" (statue rocks). There would be wide scope here for the imaginative genius who has given the nomenclature to the rocks in the Garden of the Gods. All this time it was raining in fierce showers, and we rounded the point of the bay east of Karnah in the face of one of them. A number of deer were seen quietly grazing in the valleys. A fresh wind came up from the south, and we went dashing up the bay, with the foam flying from the bow of the boat, and a boiling white wake behind us. We landed on a sandy beach near the head of the bay. While the tent was being pitched and the boat hauled out of the water a school of white whales ("kahkok-tah") came puffing and sporting into the cove, and Koomenahpik immediately went out in his kayak, which we had in tow, after them. He remained out for an hour, but as the

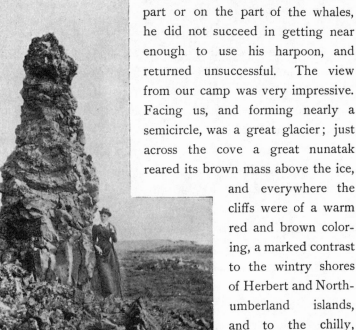

result of cautiousness, either on his part or on the part of the whales, he did not succeed in getting near enough to use his harpoon, and returned unsuccessful. The view from our camp was very impressive. Facing us, and forming nearly a semicircle, was a great glacier; just across the cove a great nunatak reared its brown mass above the ice, and everywhere the cliffs were of a warm red and brown coloring, a marked contrast to the wintry shores of Herbert and Northumberland islands, and to the chilly, gray sandstone cliffs of Karnah. Our tent was pitched just above

Pillar of Sandstone.

high-water mark beside a little stream whose banks were actually yellow with Arctic poppies.

The heavy showers continued through the night, and we waited until noon of the 11th for them to cease. Verhoeff was out after specimens until after midnight, and then, returning, slept in the boat.   He left us at this point to join Gibson in Tooktoo Valley.   Crossing over to the eastern side of the bay, we found a beautiful rock-protected cove, with a stream flowing into it from a valley above.   While Mr. Peary climbed to the top of a rock to obtain some bearings, I took my rifle and started up the valley in search of deer.   In a short time I had shot two.   One of them I brought down at long range while he was running at full speed.   As this day was the anniversary of our wedding, we celebrated it mildly with a milk punch and fried liver from the deer which I had shot.   Here, midway between the Arctic Circle and the Pole, we were in a veritable garden spot.   Vines and plants and flowers run and grow in luxuriant abundance over and in the crevices of the rocks.   The stream which empties into the cove comes from a beautiful mirror-like lake set in a grassy meadow only a short distance up the valley, and over the protecting ledge to the west come the continuous thunder and groanings of the great glacier.

Continuing our exploration, we arrived, through wind, snow, and rain, at the precipitous island which lies near the eastern extremity of the gulf.   Here, in the angle of the island and a huge glacier, in which it was partially buried, we pitched the tent, though not without protest from the natives, who said that the waves from an iceberg breaking off the glacier might smash the boat and swamp the camp.   While we were at dinner Koomenahpik raised the alarm of " kahlillowah," and looking out we saw two narwhal among the bergs, a large one and a small one. We immediately pulled out for the animals. As we approached, the larger of the two disappeared, but we were able to get near enough to the other one for me to put a bullet through its head; then Koomenahpik drove a harpoon into its back, and after a short struggle we had it in tow for the camp. The next morning we found my prize high and dry on the rocks, a great mottled, misshapen mass of flesh, with a gleaming ivory horn, straight as an arrow, and almost as sharp as a stiletto, projecting straight out from its nose. It was a wonderful sight to me, who never before had seen the narwhal, the fabled ancestor of the unicorn. I could not gaze at it sufficiently.

When we started off again, in the afternoon of August 14th, our boat was loaded down almost to the gunwales with our trophies of narwhal and reindeer, the tents, and other equipment. The morning's promise of pleasant weather had not been fulfilled. Heavy black clouds were gathering thick and fast, and by the time we had reached the southern end of the island it was raining steadily. As we ran out from the lee of the island the full force of the now furious northeast gale struck us, and we were pelted mercilessly with sheets of water. It was a wild scene, with the sullen, spectral glare of the great glaciers north and east of us beneath the pall of black clouds, the wind howling over us as if it would pick us bodily out of the water, and the black cliffs at the mouth of Academy Bay, our destination, mere shadows, felt rather than seen through the rain

full twenty miles to the south. The gulf was full of great
bergs and masses of hard blue ice, the outflow from the glaciers, through the mazes of which we were obliged to pick
our way; yet they were our friends, for they kept the water
smooth in spite of the raging wind, and gave us now and then
a shelter, behind which we could stop for a few moments and
catch our breath before striking out again into the furious
blast.   Fortunately, the wind was partly in our favor; in
spite of our tortuous course we made rapid progress, and in
four hours were abreast of the group of islands down in the
southeast corner of the gulf, which we had visited in April
during our sledge trip.   From here to Tawanah's igloo at
the mouth of the bay was the critical part of our voyage.
This distance was entirely free of ice, and though only five or
six miles in width, the force of the wind was such that the
whitecaps were rushing madly across it as we came out from
under the shelter of the islands.   With just a bit of the foresail up to enable the boat to run away from the waves, and
two oars ready to be dropped instantly into the rowlocks, in
case of necessity, we dashed madly along, with every now and
then the top of a wave coming in over the stern of the boat,
and striking Mr. Peary and myself in the back with a resound-
ing whack.   More than once my teeth involuntarily closed
more firmly as I saw a mad white crest rushing down upon
us, but our little craft rode the waves like a duck, and we
finally shot under the lee of the point at Tawanah's igloo.
As the boat sped along through the placid water and the sail

flapped against the mast in the eddy of wind under the point, every one breathed a sigh of relief.

In spite of the fury of the storm out in the gulf, here in the bay under the steep shore everything was calm and quiet. The mast and sail were taken down and the oars run out, and our native crew settled down to work again, glad to warm themselves by exercise. Suddenly, however, the wind, with the perverseness common to winds in these Arctic regions, came rushing out of the bay, meeting us full in the face, and making it almost impossible for the men to make head against it. But Mr. Peary spurred them on, and by hugging the shore we succeeded, with the aid of the tide, in reaching a little island about half-way up the bay, opposite which, despite the high waves, we effected a landing. We had the utmost difficulty in setting up our tent, but we at last got the better of the hurricane by securing the bottom of the tent all around with huge stones.

Never before had I understood the power of the wind. To add to its terrifying effect, it did not blow steadily now, as when it first commenced, but came in frightful gusts with intervals of calm between. For perhaps a minute or two it would be absolutely still, the black cliffs across the bay would loom up in perfect distinctness, and every intonation of the waves, dashing upon the rocks, could be heard; then a rushing white wall would spring into view around the bend of the bay a mile or so above us, an ominous murmur would be heard, rapidly increasing in volume and intensity, until, with a roar,

THE MELVILLE GLACIER—INGLEFIELD GULF.

the Arctic blast was upon us, literally cutting the tops off the waves and hurling them in solid masses of water far up the cliffs. The icebergs went tearing out of the bay like ships in a ten-knot breeze. A number of these bergs sailed in toward our little island, and, grounding at the upper end of the channel, formed a complete breakwater. When the wild gusts struck these great bergs they rocked and groaned, flung themselves at each other with thunderous crash, reeling backward shattered and split from the shock, while all the time the waves dashed against their outer faces, climbed in white jets clear to their tops, and fell in intermittent cataracts into the waters of our little harbor. It seemed as if we were at the very gates of the Hyperborean Inferno. All night long this struggle continued, the flying spray from the iceberg breakwater dashing against the tent, drenching it and all its contents. Mr. Peary and Matt spent the greater part of the night in holding up the tent-poles.

By morning the storm had exhausted its fury, and we were on our journey once more. But heavy weather soon set in again, and a disagreeable drizzle continued throughout the night and the greater part of the following day. We made a bee-line diagonally across the gulf to Karnah, the castellated cliffs of which could just be discerned through the gray mist which hung low over the water. Head winds and a contrary flood-tide made our progress slow, and it was only after a long and weary day of hard work for the men at the oars, and of wet and cold and cramp for those in the stern of the boat,

13*

that we touched the northern shore a few miles above Karnah, where we gladly availed ourselves of the opportunity to jump out and stretch our stiffened limbs. It was our intention to camp here for the night, but after the refreshing effects of a hot dinner, with ample draughts of tea, every one felt so much better, although thoroughly tired out, that we determined to push on to Redcliffe. As we neared Cape Cleveland the wind blew a gale, but now, for the first time, it was in our favor, and Mr. Peary ordered up both sails. Under Matt's skilful guidance we went flying past the cliffs for the mouth of McCormick Bay, dodging the hard blue lumps of ice, some of which could not be seen until we were almost upon them, frightening a herd of walrus into which we dashed unexpectedly, and then at last whirled round the point at Cape Cleveland into an eddy of quiet wind and water. Scarcely had we rounded the Cape, however, when Mr. Peary's eye saw another one of those white squalls rushing down upon us from Tooktoo Valley, and there was just time to get the masts and sails down, and the men to the oars with feet braced against the seats and backs straining to the bending ash-blades, when the squall was upon us. The wind tore off the tops of the waves and dashed them in our faces until it was impossible to see. When the gusts were at their height the men could only hold their own and prevent the boat from being blown backward out into the sound, while in the intervals between they managed to gain a little, and in this way we crept along inch by inch toward the sheltered beach on which we had landed from

the " Kite " a week before.   Suddenly, just as we came abreast
of the place where a still remaining portion of the ice-foot
formed an ugly overhanging shelf, under which the waves
broke furiously, Kulutingwah's oar snapped short off, and
Kulutingwah himself, with a wild cry, pitched backward into
the bottom of the boat.   In the momentary confusion which
followed, the boat began drifting rapidly under the shelf, when
Mr. Peary seized the oar of the man nearest him and urged
every one to his utmost, at the same time shouting to Kulutingwah to jump for the bow of the boat and throw the grapnel out.   With understanding quickened by fear, the Eskimo
carried out the order almost as soon as it was uttered, and
with all still tugging at the oars to ease the strain upon the
anchor-rope, the boat settled slowly back inch by inch, until
finally she stopped so near the wicked blue shelf of ice that I
could touch it with my hand.   This respite gave us a chance
to recover our breath, and enabled Mr. Peary to make a change
in the disposition of the men.   In the intervals between the
gusts the oars slowly and painfully worked the boat ahead,
and before the next squall struck us the grapnel was thrown
over, and every one crouched low in the boat, to present as
little surface as possible to the wind.   In this way, with the
woman Armah crying and screaming in the bottom of the
boat, and the faces of the men a dingy white, we at last
reached the coveted beach.   So deafening was the roar of the
wind that we could hardly hear each other's voices.   Leaving
Kulutingwah to watch the boat, we made our way to Redcliffe.

CHAPTER XIX

FAREWELL TO GREENLAND

Alarm about Mr. Verhoeff — A Search Instituted — Alone with Matt and the Native Women — No News — Return of the Search-parties — Poor Verhoeff — Packing up — I play Lady Bountiful — Pennsylvania's Gifts to the Natives — Farewell to Redcliffe — Fossil-hunting at Atanekerdluk — Godhavn revisited — Godthaab — Eskimo Kayakers — Fire-swept St. John's — Arrival at Philadelphia — Home again.

Thursday, August 18. When we rejoined our men at midnight we learned from Dr. Cook that Verhoeff, who left us at Bowdoin Bay, had not yet returned, and that Gibson and Mr. Bryant, the second in command of Professor Heilprin's party, were in Five-Glacier Valley searching for him. Verhoeff, after having joined Gibson, left him at the valley for a further search after minerals, and his last words were, " If I am not here don't be worried; I may be gone till Tuesday or Wednesday."

Before retiring Mr. Peary sent a note on board the " Kite," informing Professor Heilprin of our return, and stating that we should be ready to say farewell to Redcliffe the next day. Soon after breakfast this morning Mr. Peary began getting the boxes and barrels of specimens ready for shipment, while

I took charge of the household effects, provisions, etc. While we were thus occupied our boat was seen coming from Five-Glacier Valley. When it had approached near enough for us to distinguish the occupants, we saw there were only two white men in it — Gibson and Mr. Bryant. Gibson told us that they had waited at the appointed place until their provisions gave out, and then had taken a scout up the valley for some distance, but had seen no sign of Verhoeff. They left a note for him, intending to return for a further search.

We now began to feel grave apprehensions regarding the missing man, and a vigorous search was immediately determined upon. Mr. Peary set to work to provision the boat; then, summoning about him all the native men, who are as expert as our Indians in following a trail, he told them that they must go with him to Five-Glacier Valley and look for Verhoeff, promising a rifle and ammunition to the man who should first discover him. Professor Heilprin then suggested that while Mr. Peary and his men went up McCormick Bay to the mouth of the valley, he and his party should go round in the "Kite" to the head of the valley in Robertson Bay; and it was so decided, and the Eskimos were divided between the two parties. I remained at Redcliffe with Matt and the native women and children.

At two o'clock the search-parties left, and I turned my attention once more to packing. The women stood around me, devoured with curiosity as to what I would do with all these things, and plying me with questions as to whose husband would win the coveted prize. They would not believe that I did not know, because I had known that Mr. Peary and Astrup would return from the inland ice.

Friday, August 19. The day is not a promising one; dark clouds are gathering and the air seems oppressive. I trust that the search-parties will find Mr. Verhoeff to-day, for he must be running short of provisions by this time. We calculated that what he had could by economizing be made to last him through Wednesday, and to-day is Friday. There is no sign of boat or ship.

Most of our provisions are stowed away on the "Kite," among them all the fresh meat; in the excitement we forgot to get any out for our use, and to-day we are living on crackers and coffee.

Sunday, August 21. When this morning's fog lifted at noon, the "Kite" was seen off Five-Glacier Valley. All day yesterday we watched for her and waited for some news, but heard and saw nothing. Seeing the vessel, I supposed of course that Verhoeff had been found, and the "Kite" had gone round to the valley to pick up the rest of the party.

After hours of watching we saw the "Kite" get up steam and head down the bay toward Redcliffe, and late in the afternoon she stopped opposite our house, and the professor came off to me in a boat, only to bring the distressing news that nothing had been seen or heard of Verhoeff. Mr. Peary was then exploring the shore from the mouth of the valley around Cairn Point to the head of Robertson Bay, where it was intended that the "Kite" should join him. Another party were making thorough search through the valley. After leaving me some provisions the "Kite" continued on her way to Robertson Bay.

Tuesday, August 23. We have had no tidings from the search-parties since the "Kite" left us Sunday evening. I am very much afraid that we shall never see our lost companion alive again. The weather since he has been in the field has been exceptionally cold, raw, and wet, and he was clothed very lightly; besides, his food must have given out some days ago. The natives all agree that no one could have slept without shelter in the furious gales which we have had lately, clothed as lightly as Verhoeff was; and as they have the experience which we lack, I cannot help feeling that there is truth in what they say, so to-night I go to bed with a heavy heart. With the dark winter night passed in safety and comfort, and the long sledge journey accomplished successfully, it seems sad indeed that we should now, on the eve of our departure, meet with so great a loss.

Wednesday, August 24. About two o'clock this morning Mané came running in to me with the news that the ship was coming, and I at once went out on the beach to await her. In half an hour she dropped anchor, and Mr. Peary, with the other members of our party, came ashore bringing the sad tidings that Verhoeff's footprints had been found and traced upon a great glacier which was cut by numberless wicked-looking crevasses, and there lost. After searching the glacier in every direction without success, there was no doubt left that poor Verhoeff had lost his life in an effort to cross the ice-stream. Mr. Peary cached enough provisions to last one man a year, at Cairn Point, in case Verhoeff should, in some miraculous way, return after the " Kite's " departure.

It was with a feeling akin to homesickness that I took the pictures and ornaments from the walls of our little room, pulled down the curtains from the windows and bed, had Matt pack the books and nail them up, sorted the things on the bed, and packed those I wanted to keep. The tins and cooking utensils I put on the stone and turf wall just outside of my room previous to distributing them among the natives.

My trunk packed and removed, the carpet up and the curtains down, the improvised bookcase taken to pieces, and it was hard to imagine that this dismantled room had once been as snug and comfortable as any boudoir in the world. Could the walls talk they would tell of some very pleasant hours spent there by the members of the North Greenland Expedition of 1891–92, and of many months of real solid comfort and happiness enjoyed by the woman who, when she left home and friends, was told over and over again that she must expect to endure all kinds of hardships, to suffer agony from that dreaded Arctic enemy, scurvy, etc.

I next turned my attention to the various articles put aside for the Eskimos, and after sorting them over I called all the women in the settlement to me, and stood them in a row. There were nine among them, including the two brides (mere children), Tookymingwah, wife of Kookoo, and Tungwingwah, wife of Kulutingwah. When they had grasped the idea that I was about to present them with these things they fairly danced with joy, shouting to their husbands, and laughing and talking with each other. I took care that Mané and M'gipsu, who had been with us constantly sewing and curing skins, should have the more desirable articles, while the others shared equally. After the distribution the professor, with a few members of his party, rowed off to the "Kite," and in a short time returned with their boat laden with pots, kettles, knives, scissors, thimbles, and needles for the women, and long ash-poles, timber cut suitable for kayaks, lances, saws, gimlets, knives, etc.—in fact, everything in the hardware and lumber line that could be of any possible use to the men. Then all the natives were collected on the beach and the different

Receiving Gifts of Charity.

articles distributed among them. I know if the good Pennsylvanians who sent these gifts could have seen the pleasure these poor natives derived from them they would have felt amply repaid.

We spent a couple of hours in taking photographs of the natives, their tupics, our poor little abandoned house and its surroundings, and then bade farewell to Redcliffe. It had been my home for thirteen months—some of them had seemed more than twice as long as any ordinary month—and I felt sorry to leave it to the mercy of wind and weather and Eskimo. Mané asked me if she might pitch her tupic in my room, saying it would be so nice and dry, and the wind could not strike it and blow it over; then, too, no matter how cold it might be, her ikkimer would be sufficient to heat it comfortably. I told her she might do so, but she must take good care of the house and not allow others to destroy anything about it, until the return of the next sun, when, if we did not come back, it should belong to Ikwa and herself to do with as they wished.

It was about noon when I left the settlement with the last boat-load, and as soon as we were safely on board the "Kite" the work of raising the anchor was begun. In the meantime Ikwa and Kyo in their kayaks were paddling round and round the "Kite," calling to us their last good-byes. Ikwa asked if he might come aboard just once more, and on permission being granted, he immediately climbed over the side and jumped on deck. Some one took a fancy to his kayak paddle, which had been broken and mended, as only an Eskimo can mend, in at least a dozen different places, and gave him an old sledge-runner for it.   When the time came for the Old Pirate to leave us all of us felt badly, and when he said " Gooby," with his peculiar accent, his eyes filled and he choked. After this he would not turn his head in our direction, and only waved his hand in answer to our good-byes.   His picture, as he paddled himself with the sledge-runner, curved at both ends, to the shore, will never fade from my memory.

As the " Kite " steamed slowly down the bay the natives ran along the beach, shouting to us and waving their hands, Kulutingwah bringing up the rear with a torn American flag attached to a pole, which he waved frantically to the imminent danger of those near him.   I could not help thinking, Have these poor ignorant people, who are absolutely isolated from the rest of humanity, really benefited by their intercourse with us, or have we only opened their eyes to their destitute condition?   I hope the latter is not the case, for a happier, merrier set of people I have never seen; no thought beyond the present, and no care beyond that of getting enough to eat and to wear.   As we steamed down the bay we turned our eyes on the red cliffs, and when they faded from view Cape Cleveland and Herbert and Northumberland Islands were the only familiar landmarks left in sight.   On these we gazed with the feeling that we were looking our last upon the scene. The old Cape, especially, seemed very near and dear to me; twice it had sheltered and protected me from the fury of an

Arctic gale — once in the winter when Mr. Peary and the doctor had gone to rescue " Jack," my pet Newfoundland, from
its precipitous cliffs, and the second time only a few days
ago, when we returned from our venturesome boat journey
up Inglefield Gulf.

Our home journey was almost wholly devoid of incident.
Melville Bay, smooth as glass, had lost its terrors, and we
steamed through it almost without hindrance. We reached
Atanekerdluk, in the Waigatt, on August 29th, and there
spent a delightful and profitable day in collecting fossils
among the "leaf beds" which have been made famous to
geologists. The following morning we arrived at Godhavn,
where once more we enjoyed the kind hospitality of Inspector
and Mrs. Anderssen, and the pleasing attentions of a daughter
who had only recently returned from Denmark. The same
friendly reception awaited us at Godthaab, the capital of the
Southern Inspectorate of Greenland, where the honors of hospitality were divided between Inspector and Mrs. Fencker and
Governor and Mrs. Baumann. It was here that Nansen descended from the ice-cap after his memorable journey across
the Land of Desolation and passed a long, weary winter of
waiting.

The Eskimos of this region have the reputation of being
the most expert kayakers in the whole of Greenland, and
we were witness to some of their most remarkable feats,
such as describing a complete revolution through the water,

SADDLE MOUNTAIN. — GODTHAAB.

and crossing one another at right angles, one canoe shooting over the bow of the other. These performances, which are

Sports of the Kayakers.— Overturning.

said to have been at one time common with all the west-coast Eskimos, are rapidly becoming a lost art, and it has even been doubted if they took place at all.

Our kind friends were so pressing in their attentions that it was not without regret that we were forced to bid adieu to their hospitable homes and a last farewell to the Greenland shores. After a rather tempestuous voyage we arrived at St. John's, Newfoundland, on September 11th, to find a scene of desolation, and wreck and

Kayaker Overturned.

14

ruin running in the path of the recent conflagration. The fire had broken out two days after the departure of the " Kite " on her last mission of good-will, and this was the first intimation that any of us had had of the catastrophe. Shaping our course southward, we arrived, after an uneventful voyage, at our port of destination, Philadelphia, where on the 24th, amid a chorus of cheers and hurrahs, and the tooting of innumerable horns and whistles, we received the congratulations of the multitude that had assembled to await our arrival.

I returned in the best of health, much stronger than when I left sixteen months before. The journey was a thoroughly enjoyable one. There were some drawbacks, it is true, but we meet with them everywhere, and were it not for the sad loss of Mr. Verhoeff, I should not have a single regret.

CHAPTER XX

GREENLAND REVISITED

Along the Labrador Coast—Strange Passengers on the " Falcon "—Holstein-borg and Godhavn—The Quickest Passage of Melville Bay—Meeting with Old Friends—No Tidings of Verhoeff—Establishing Ourselves at Bowdoin Bay—Deaths among the Eskimos—A Rich Walrus Hunt—Smith Sound and the Northern Ice-pack—Polaris House—Departure of the " Falcon."

Anniversary Lodge, Bowdoin Bay, Greenland, August 20, 1893. The reader who has followed me through my Arctic experiences of 1891–92 may be interested to know how we found our Eskimo friends upon our return to them after an absence of nearly a year.

On July 8 the steamship " Falcon," carrying north the members of Mr. Peary's new Arctic expedition, left Portland, and headed for St. John's, where we landed on the 13th. We had with us a conglomerate cargo, including, in addition to the ordinary paraphernalia of an Arctic expedition, eight little Mexican burros or donkeys, two St. Bernard dogs, the Eskimo dogs which Mr. Peary had brought down from Greenland, and numerous homing pigeons, kindly presented to us by friends interested in the expedition. At St. John's we added a few

Newfoundland dogs, and then proceeded north along the Labrador coast, touching at several of the missionary stations, where
we obtained about thirty dogs from the Eskimos.   It was a
pitiable sight to see how famished these poor Moravian missionaries were for news from the old as well as the new country.   They have direct mail communication with Europe only
once a year.

I was told that although they have only three months in
the year when frost is out of the ground, yet they all cultivate
small gardens, and the most delicious dish of stewed rhubarb
that I ever tasted was prepared from a bundle sent to me by
one of the missionaries.   It was interesting to note that while
the appearance of the Labrador Eskimos is very similar to that
of the natives of South Greenland, yet their mode of dress
is different in both pattern and material.   The undershirts,
instead of being made of the skins of birds, are made of blanketing, and instead of being the same length back and front,
are fashioned with a long tail; over this is worn a garment of
the same pattern, made of drilling.   The trousers are also of
woven material.   Of course this was their summer costume.
The women all wore blanket skirts, and had woolen shawls
about their shoulders.

After following the coast of Labrador for ten days, we
headed across Davis Strait for Holsteinborg, on the Greenland shore.   It took us about twelve hours to steam through
the stream of ice which was flowing southward, but only once
did the " Falcon " have to go astern in order to move a pan of

ice and make a passageway for herself. Steadily she steamed on, butting against the cakes and floes until her timbers quivered and creaked. At last we were in clear water again, and then our vessel fairly bounded over the waves.

Arrived at Holsteinborg, we found a pretty, clean little village. There are more wooden houses here than at Godhavn, and altogether the place looks more thrifty. We found the governor absent, but the assistant governor, a young Danish officer who spoke a little English, did the honors, and he procured twenty-three dogs from the natives for us. Among other attentions, he sent to me a basket of radishes, fresh from his garden.

Business completed, the " Falcon " steamed north for Godhavn. On our arrival at this little hamlet we found everything apparently unchanged, but, to our great disappointment, our pilot informed us that Inspector Anderssen was absent on a tour of inspection, accompanied by his daughter, and that Governor Joergensen and family had gone to Denmark. We found Mrs. Anderssen as rosy-cheeked and as youthful as when we first saw her. She made our visit very pleasant, rounding it off with one of her delightful little dinners on the evening of our departure. We requited her hospitality by presenting her with various kinds of fruit — pineapples, lemons, oranges, and a watermelon. The natives expressed great pleasure on seeing us, and old Frederick, who had accompanied Mr. Peary on the ice in 1886, after shaking hands with me, said, " Very gude, you look all samee," rubbing his hands over his face and

14*

then pointing to mine to show me that I had not changed in looks since last he saw me.

Our next stopping-place was Upernavik, where we remained just long enough to pick up a few dogs, after which we put in at Tassiusak, the most northerly inhabited spot in the world belonging to any government. This place boasts of but a single wooden house. We here still further increased our stock of dogs, and then left. The next day we revisited the Duck Islands, but this year the sport did not compare with that of two years ago, when the birds were so plentiful that one could hardly walk without fear of stepping on them. This year it was a month later in the season, and not only were the young ducks hatched, but the old mother ducks were out teaching the ducklings to swim, and the islands consequently were all but deserted. I devoted my time to the gathering of down for the bedding in our Arctic home, and secured about thirty pounds.

We now headed for the ever-dreaded Melville Bay, my first experience with which I shall never forget. We were then three weeks in crossing, and it was during that time that Mr. Peary had the misfortune to have his leg broken. This time everything looked favorable; we had no fog, and there was no ice in sight from the crow's nest. Captain Bartlett was determined to break the record in the crossing of this water—thirty-six hours—on this his first voyage to the Arctic regions. In twenty-four hours and fifty minutes we reached the Eskimo settlement at Cape York, Melville Bay behind us and still no ice to be seen.

At this settlement, where formerly so many natives lived, we found only three families, all of them strange to us; they could tell us nothing about our acquaintances in the tribe, not having seen any of the inhabitants to the north of them since the time we left McCormick Bay. We pushed on along the Greenland coast until we rounded Cape Parry, and then steamed into Barden Bay, stopping at the Eskimo village of Netchiolumy. Here, too, instead of finding about sixty natives, as was the case a year ago, we found only two families. Mr. Peary with two men went ashore at once, and before their boat reached the land I heard one of the natives shout "Chimo Peary," and saw him dance up and down for joy. On his return Mr. Peary informed me that the natives were Keshu, *alias* the Smiler, and Myah, the White Man, with their families. They were wild with delight, and begged to be allowed to accompany us to the site of our new house and pitch their tents beside it. They were stowed with all their belongings into Mr. Peary's boat, and in a short time both families with their houses and their chattels were on board the "Falcon." They gave us all the news and gossip of the tribe. Naturally, we first questioned them about our lost companion, Mr. Verhoeff. There never was a doubt in our minds that Mr. Verhoeff lost his life in crossing the glacier at the head of Robertson Bay; but his friends at home took a different view of the matter, and were confident that we would find him alive and well. These natives say that nothing has been seen or heard of him, and they hesitate to speak of him, as they never speak of their

dead.   Mr. Peary thought perhaps some article of his clothing
had been found by the Eskimos that might throw some light
on the disappearance of our unfortunate associate; but nothing
whatever has been found.   We next inquired about our Eskimo friends, and were grieved to hear that the little five-
year-old, bright-eyed, mischievous Anadore, daughter of our
henchman Ikwa and his wife Mané, had died in the early
spring.   We learned that Redcliffe House had been destroyed
by a few of the natives, led on by the famous angekok, Kyoa-
ahpadu, and that he had also destroyed the provisions which
were cached at Cairn Point by Mr. Peary.

We arrived at our destination, at the head of Bowdoin Bay,
on August 3d, without any difficulty, the ice having almost
completely left the bay and sound.   The Sculptured Cliffs of
Karnah, forming the cape of Bowdoin Bay, stood out sharp
and clear in the early morning sunlight, while the towering red

The Cliffs of Karnah.

Castle Cliffs frowned down upon the bay from the opposite cape.

The site selected for our new home is only a few feet from where we pitched our tent last year when engaged in the exploration of Inglefield Gulf, and where, amidst a furious rainstorm, we celebrated our wedding anniversary. As we shall celebrate at least two more such anniversaries here, we have christened our new home "Anniversary Lodge." The great cliff which mounts guard over us Mr. Peary has named Mt. Bartlett, in honor of our gallant young skipper, Captain Harry Bartlett, of St. John's. Our snug and picturesque harbor is to be known as Falcon Harbor, named after the little bark which brought us here in safety, and which is the first ship to anchor in these waters.

The day after we dropped anchor in Falcon Harbor we were visited by five of our former Eskimo acquaintances, who had paddled at least twenty-five or thirty miles in their kayaks on seeing the ship pass their settlement. Two of them, Kulutingwah and Annowkah, were residents of Redcliffe, and it really seemed like meeting old neighbors, although I must confess that they appear even dirtier than they did a year ago. Annowkah told me that his wife, M'gipsu, who was our most skilful seamstress, was ill; but it is impossible to get these people to talk much about their sick, and so I was unable to find out what really ailed the poor woman.

Our Eskimos stayed with us a few days and assisted us in landing our supplies. They were vastly amused at the burros,

which they persist in calling "big dogs"; and I can hardly
blame them, for my St. Bernard dog is almost as large and
tall as some of these little animals.   After the provisions were
all ashore, each native took a load of about fifty pounds on his
back and carried it to the ice-cap; but this was the last straw,
and every man decided that he really must return to his
family at once.

On August 12, the work on the house being well advanced,
Mr. Peary decided to make a trip after walrus for dog-food,
intending to proceed as far as Smith Sound, if possible.   It
takes quite a little pile of meat to feed eighty-three Eskimo
dogs.   Accompanied by the two natives, Keshu and Myah,
we started for Karnah, the nearest settlement, where we had
intended to pick up one or two additional hunters; but on
reaching the place we were shocked to hear that M'gipsu had
died "two sleeps ago."   Mr. Peary went to Annowkah's tent,
and there sat the bereaved husband, with his sealskin hood
pulled over his head, looking straight before him, saying nothing and doing nothing, apparently knowing nothing of what
was going on about him.   It is the custom with these people
to act in this way for a certain length of time after a death,
and then they desert the hut or tent in which the death has
taken place, and it is never again occupied.   M'gipsu's little
six-year-old boy, whose father died when he was very small,
also sat in the tent all huddled up in one corner.   Poor little
fellow! I do not know what will become of him now, for it is an
open secret that his stepfather, Annowkah, does not like him.

As we proceeded up the sound we saw the cakes of ice thickly sprinkled with walrus, which had come out of the water and were taking a sun-bath. The boats were lowered, and the men started after them. In a few hours we had twenty-four of the monsters on board. Their average weight was estimated at not less than fifteen hundred pounds. There were several cold baths taken by the hunters, and some narrow escapes, but nothing serious occurred, and we continued on our course, heading for Cape Alexander. Once around the cape, we steamed half-way across the sound toward Cape Sabine, where we were stopped by the ice-pack, which stretched in an unbroken plain as far as we could see. Turning back, we visited the site of the Polaris House, where a portion of Captain Hall's party wintered after the " Polaris " was wrecked. We picked up a number of souvenirs in the shape of bolts, hooks, hinges, even buttons and leaves from books. A quantity of rope was found on the border of a little pond just back of where the house stood, and it seemed to be in a state of perfect preservation. We also stopped at Littleton Island, and on the adjoining McGary Island some of the party indulged in a little shooting. A few ducks and guillemots were shot; four additional walrus and an oogzook seal were also obtained in this vicinity. The weather then became thick and a strong wind sprang up, which put an end to the sport.

All night we steamed toward Hakluyt Island, but on reaching it we could not make a landing on account of the gale. We lay in the shelter of the cliffs of Northumberland, and

when the storm abated steamed along its shore, and, crossing
Whale Sound, entered Olrich's Bay, the scenery of which sur-
passes that of any of the other Greenland bays that I have
seen. Our party scattered at once in search of reindeer,
which we were told were numerous here, and in a few hours
we had seventeen on board ship.

Our house is up, and promises to be very cozy. The good
ship "Falcon" sails for home to-morrow, taking with her the
last messages which we can send our dear ones for some time.

Everything points to the success which Mr. Peary hopes
for. What the future will bring, however, no one can tell.

THE GREAT WHITE JOURNEY

FROM McCORMICK BAY
TO THE NORTHERN SHORE OF GREENLAND
AND RETURN

BY

ROBERT E. PEARY

THE GREAT WHITE JOURNEY.

According to my program, the 1st of May was to be the time for the start on the inland ice, and on the 28th of April, Astrup, Gibson, Dr. Cook, and the native men then at Redcliffe left with the last load of supplies for the head of McCormick Bay. The natives were to return after helping the boys carry the supplies to the top of the bluff; the boys themselves were to push forward with the work until I joined them. This I did on the 3d of May. When I left Redcliffe the number of natives there had dwindled very materially; some drawn away to the seal-hunt, but more driven away by their superstitious feeling in regard to my going upon the great ice. We had the most exceptionally fine weather all through April, but on the very night that I reached the head of the bay a sullen sky over the ice-cap betokened a change. From this night until the morning of the 6th of August, when Astrup and myself clambered down the flower-strewn bluffs again, my couch was the frozen surface of the inland ice, and my canopy the blue sky.

The first two weeks after leaving the little house upon the shores of McCormick Bay were occupied in transporting the supplies—which at various times during the preceding month

had been carried by the members of my party and helping
natives to the crest of the bluffs at the head of the bay — to
the edge of the true inland ice, some miles distant, and then
in dragging them over and among the succession of the great
domes of ice which extend inward some fifteen miles to the
gradual slope of the vast interior snow-plain.   One or two
snow-storms and the constant violent wind rushing down from
the interior to the shore, combined with the difficulties of the
road and the constant annoyance from our team of twenty
savage and powerful Eskimo dogs, entirely unaccustomed to
us and to our methods, made these two weeks a time of un-
remitting and arduous labor for myself.   The only pleasant
break in this work was the occurrence of my own birthday,
and the unexpected appearance from among the medical
stores, in charge of Dr. Cook, of a little box from the hands
of the dear one left behind, containing a bottle of Château
Yquem, a wine endeared to both of us by many delightful
associations, a cake, and a note containing birthday wishes for
success and continued health.   Once on the true ice-cap, two
good marches brought us to the divide, from which, as from
the ridge of a great white-roofed house, the ice-cap slopes
north to the shores of Kane Basin and historic Renssellaer
Harbor, where Kane and his little party passed so many Arctic months, and southward to the shores of Whale Sound and
our own little home.   From this divide we had a slight descent in our favor, and we kept on from the edge of the basin
of the Humboldt Glacier, where the great mass of the inland

SAILING OVER THE INLAND ICE.

ice, like very cold molasses, hollows itself slowly down to the mighty glacier itself. Here the fiercest storm that we had encountered thus far burst upon us, and for three days we were confined to our snow shelter, getting out as best we could in occasional lulls in the storm to secure loose dogs and endeavor to protect the loads upon the sledges from their ravages. In this we were fairly successful, though we did not succeed in preventing them from devouring some six pounds of cranberry jam, and eating the foot off Gibson's sleeping-bag. This storm over, we were not again troubled by really violent storms during our northward march.

On the 24th of May Dr. Cook and Gibson, who had formed our supporting party, left us to return to Redcliffe, leaving Astrup as my sole companion for the remainder of the journey. On the last day of May, from the dazzling surface of the ice-cap we looked down into the basin of the Petermann Glacier—the grandest amphitheater of snow and ragged ice that human eye has ever seen, walled in the distance by a Titan dam of black mountains, and all lit by the yellow midnight sunlight. Still keeping on to the northward, navigating the ice as does the mariner the sea along an unknown coast, we were befogged for two or three days in clouds and mists which prevented us from seeing to any distance. As a result, we approached too near the mountains of the coast, and got entangled in the rough ice and crevasses of the Sherard Osborne Glacier system. Here we lost twelve or fourteen days in our efforts to get back to the smooth, unbroken snow-cap

15

of the interior. Once there, we continued our march, always northeastward, till on the 27th of June I discerned black mountain-summits rising above the horizon of the ice-cap,

directly ahead of us. Then the northwest entrance of a fjord came into view, and we could trace its course southeasterly just beyond the nearer mountains of the land north and northeast. I changed my course to east, when I was soon confronted by the land and the fjord beyond. Then I turned to the southeast, and traveled in that direction until the 1st of July, when we, after fifty-seven days of journeying over a barren waste of snow, stepped upon the rocks of a strange new land, lying red-brown in the sunlight, and dotted with snow-drifts here and there. The murmur of rushing streams, the roar of leaping cataracts from the ice-cap, and the song of snow-buntings made the air musical. Leaving the sledge and our supplies at

The Land beyond the Ice.

the very edge of the rocks, leading our dogs, and with a few days' supplies upon our backs, Astrup and myself started on over this strange land, bound for the coast, which we knew could not be far distant. Four days of the hardest traveling, over sharp stones of all sizes, through drifts of snow and across rushing torrents, and we came out at last upon the summit of a towering cliff, about 3500 feet high, now known as Navy Cliff, from which we overlooked the great and hitherto undiscovered Independence Bay.

Before us stretched new lands and waters, to which, with the explorer's prerogative, I gave names, as follows: the bay at our feet, opening into the Arctic Ocean half-way between the 81st and 82d parallels of latitude, was named Independence Bay in honor of the day, July 4th; the red-brown land beyond the fjord which had stopped our forward northward progress was called Heilprin Land; and a still more distant land beyond the entrance of a second fjord, Melville Land. The enormous glacier at our right, flowing due north into Independence Bay, received the name of Academy Glacier, and the bold rugged land beyond it, Daly Land.

It was almost impossible for us to believe that we were standing upon the northern shore of Greenland as we gazed from the summit of this bronze cliff, with the most brilliant sunshine all about us, with yellow poppies growing between the rocks around our feet, and a herd of musk-oxen in the valley behind us. Two of these animals we had killed, and their bodies were now awaiting our return for a grand feast of fresh meat. Down in that same valley I had found an old friend, a dandelion in bloom, and had seen the bullet-like flight and heard the energetic buzz of the bumble-bee.

For seven days we remained in this northern land, more than six hundred miles of pathless icy sea separating us from the nearest human being, and then we began our return march. This return march, much shorter than the upward

The Academy Glacier.

one, was uneventful and monotonous. For about two weeks we were about a mile and a half above the sea-level, literally in the clouds, and day after day, in every direction, stretched only the steel-blue line of the snow horizon. The snow was soft and light, and without our " ski," or Norwegian snow-skates, and Indian snow-shoes we should have been almost helpless in it; but at last, after passing the latitude of the Humboldt Glacier, when we were only about a mile above the sea-level, the traveling became better. The slight down-grade assisted us, and for seven days we averaged thirty miles a day, increasing our distance on each successive day, showing that both men and dogs were in perfect training, and, like the scientific athlete, had still the reserve force necessary for a grand spurt on the home stretch.

The night of the 5th to the 6th of August was an exquisitely clear and perfect one. From eight to eleven Astrup and myself and our remaining five dogs toiled up the north slope of the largest of the ice-domes between the head of McCormick Bay and the edge of the true interior ice—one to which I had given the name "Dome Mountain." As I rose

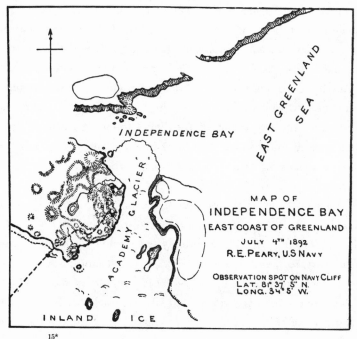

INDEPENDENCE BAY

EAST GREENLAND SEA

ACADEMY GLACIER

MAP OF
INDEPENDENCE BAY
EAST COAST OF GREENLAND
JULY 4TH 1892
R. E. PEARY. U.S NAVY

OBSERVATION SPOT ON NAVY CLIFF
LAT. 81° 37' 5" N.
LONG. 34° 5' W.

INLAND ICE

15*

over the crest of the great white mass and looked down and forward upon our course, there, some two miles away, upon the slope of the next dome, were two or three dark, irregular objects. Even as I looked at them they moved and separated, until I could count several detached bodies. They could be but one thing—men; and as there were so many of them, and as I was sure that none of the Eskimos could have been persuaded by my boys to set foot upon the inland ice, I knew in an instant that some ship was lying in the bay waiting for us. It was but a little while later, both parties descending rapidly toward each other, that we met in the depression between the two domes, and I grasped again the hand of Professor Heilprin, who had been the last to say good-by to me a year before, as I lay a cripple in my tent, and who now had come again to meet me and bring us back. It was a strange and never-to-be-forgotten meeting. In the ship lying at anchor at the very head of the bay I found the woman who had been waiting for me for three months, and two days later we were back again in the little house which had sheltered us through a year of Arctic vicissitudes.

Such, in brief, is the outline of the inland-ice journey from McCormick Bay to the northern shore of Greenland and back. Its important results are already well known, and it is not necessary to revert to them here. I will attempt, however, to give some adequate impression of the unique surroundings in which our work was done, and also to make clear the real character of this great interior ice-plateau, a natural feature so entirely different from any with which we are aquainted in better known portions of the globe that I have sometimes found it difficult to convey, even to the most cultivated minds, a really adequate conception of what the great ice-cap is like.

The terms "inland ice" and "great interior frozen sea," two of the more common names by which the region traversed by us is generally known, both suggest to the majority of people erroneous ideas. In the first place, the surface is not ice, but merely a compacted snow. The term "sea" is also a misnomer in so far as it suggests the idea of a sometime expanse of water subsequently frozen over. The only justification for the term is the unbroken and apparently infinite horizon which bounds the vision of the traveler upon its surface. Elevated as the entire region is to a height of from 4000 to 9000 feet above the sea-level, the towering mountains of the coast, which would be visible to the sailor at a distance of sixty to eighty miles, disappear beneath the landward convexity of the ice-cap by the time the traveler has penetrated fifteen or twenty miles into the interior, and then he may travel for days and weeks with no break whatever in the continuity of the sharp, steel-blue line of the horizon.

The sea has its days of towering, angry waves, of laughing, glistening white-caps, of mirror-like calm. The "frozen sea" is always the same—motionless, petrified. Around its white shield the sun circles for months in succession, never hiding his face except in storms. Once a month the pale full moon climbs above the opposite horizon, and circles with him for eight or ten days.

Sometimes, though rarely, cloud shadows drift across the white expanse, but usually the cloud phenomena are the heavy prophecies or actualities of furious storms veiling the entire sky; at other times they are merely the shadows of dainty, transparent cirrus feathers. In clearest weather the solitary traveler upon this white Sahara sees but three things outside of and beyond himself—the unbroken, white expanse of the snow, the unbroken blue expanse of the sky, and the sun. In cloudy weather all three of these may disappear.

Many a time I have found myself in cloudy weather traveling in gray space. Not only was there no object to be seen, but in the entire sphere of vision there was no difference in intensity of light. My feet and snow-shoes were sharp and clear as silhouettes, and I was sensible of contact with the snow at every step. Yet as far as my eyes gave me evidence to the contrary, I was walking upon nothing. The space between my snow-shoes was as light as the zenith. The opaque light which filled the sphere of vision might come from below as well as above. A curious mental as well as physical strain resulted from this blindness with wide-open eyes, and sometimes we were obliged to stop and await a change.

The wind is always blowing on the great ice-cap, sometimes with greater, sometimes with less violence, but the air is never quiet. When the velocity of the wind increases beyond a certain point it scoops up the loose snow, and the surface of

DRIFTED IN.

the inland ice disappears beneath a hissing white torrent of blinding drift. The thickness of this drift may be anywhere from six inches to thirty or even fifty feet, dependent upon the consistency of the snow. When the depth of the drift is not in excess of the height of the knee, its surface is as tangible and almost as sharply defined as that of a sheet of water, and its incessant dizzy rush and strident sibilation become, when long continued, as maddening as the drop, drop, drop, of water on the head in the old torture-rooms.

While traversing the inland ice our hours of marching were those corresponding to what here would be night—that is, when the sun was above the northern horizon. In our line of march I took the lead, on snow-shoes or ski as the condition of the snow demanded, setting the course by compass, or by time, and the shadow cast by my bamboo staff. The dogs, a few yards in the rear, followed my trail, and Astrup traveled on ski beside the sledge, encouraging the dogs and keeping them up to their work.

Our daily routine was as follows: When the day's march (measured sometimes by the hours we had been on the move and sometimes by the distance covered) was completed, I began sounding the snow with the light bamboo staff to which my little silken guidon was attached, until I found a place where it was firm enough to permit of blocks being cut from it. This done, the guidon-staff was erected in the snow, and at the shout of "Tima" from me, my dogs, no matter how long or how hard the day had been, would prick up their ears and come hurrying up to me until they could lie down around my feet, glad that the day's work was done.

As soon as the sledge came to a standstill I read the odometer, aneroid, and thermometer; then Astrup and myself undid the lashings, and as soon as the lines were loose Astrup took the saw-knife and began excavating for our kitchen, while I took the short steel-pointed stake to which we fastened our dogs and drove it firmly into the snow in front, and some fifty feet to leeward, of the kitchen site. I then untangled the dogs' traces, detached the animals from the sledge, and made them fast to the stake. I next got out a tin of pemmican, a can-opener, and a heavy hunting-knife, and, kneeling behind the sledge, prepared the dogs' rations, which consisted of a pound of pemmican each. I then fed the hungry creatures, standing over them meanwhile with the whip, to see that the weaker ones were not deprived of their share.

By this time Astrup had completed an excavation in the snow, about eight feet long by three feet wide and a foot and a half deep, and with the snow blocks obtained from this excavation had formed a wall a foot or a foot and a half high across one end and half-way down each side. Across this wall was put one, and sometimes both, of the ski, and over this was spread a light cotton sail, weighted down with blocks of snow. This was known as our kitchen, and at the innermost end was placed the kitchen-box, containing our milk, tea, pea-soup, Liebig's Extract, drinking-cups, can-opener, knives, spoons, and the day's rations of pemmican and bis-cuit; also the alcohol-stove and a box of matches, done up in a waterproof package.

Then, if it was Astrup's turn as cook he immediately began the preparations for dinner by lighting the alcohol-lamp and filling the boiler with snow, while I lay down in the lee of the sledge and made my notes of the day's work. If it was my turn as *chef*, as soon as the kitchen was finished I took pos-session of it, and Astrup retreated to the shelter of the sledge. While the snow was melting I wrote up my notes, Astrup usually devoting this time to rubbing vaseline into his face to repair the ravages of the sun and wind. As soon as sufficient water had been melted, two cupfuls of pea-soup were made, and this, with a half-pound lump of pemmican, formed our first course. While we were enjoying this the water for our tea was brought to the boiling-point. Pea-soup and pemmi-can finished, we each had a cupful of cold milk, and when this had disappeared the tea was made; six biscuits apiece formed our dessert.

When our luxurious repast was over, what was left of our day's allowance of alcohol was allowed to expend itself on a fresh boilerful of snow for our morning tea, while the cook made his preparations for the night by changing his footgear and tightening the drawstrings of his furs. In addition to his other duties, the cook of the day had the entire responsibility, from dinner-time until breakfast, of the dogs, and it was the first rigid regulation of the journey that he should always be so dressed that he could at a moment's notice jump from his shelter and capture a loose dog. The dogs were always fastened directly in front of the opening of the kitchen, so that the occupant, by raising his head, could see at once if his presence were needed. During the first portion of our journey this duty was an onerous one, and frequently meant a sleepless night; but later on, after several of the dogs had received some severe discipline for attempted thefts, and particularly after we had adopted the plan of muzzling them every night as soon as they had finished their dinner, we had but little trouble.

In the morning I was generally the one to waken first, and would either start the alcohol-lamp myself or else call Astrup for that purpose. Our morning meal consisted of a lump of pemmican, six biscuits, two ounces of butter, and two cups of tea each. As soon as this was finished everything was repacked on the sledge, and while Astrup was completing the lashing, I removed the dogs' muzzles, untangled their traces, and attached them to the sledge. I then read the odometer, aneroid, and thermometer, and, taking the guidon, which had waved and fluttered over the kitchen throughout our hours of rest, from its place, stepped forward, and the next march was commenced. After from four to six hours of marching we would halt for half an hour to eat our simple lunch of pemmican and give the dogs a rest, and then, after another four to six hours of traveling, halt again and repeat the already described routine.

The three sledges used on our journey were the survivors of a fleet of ten, comprising seven different styles. They consisted simply of two long, broad wooden runners curved at both ends, with standards supporting light but strong cross-bars. The largest sledge was thirteen feet long and two feet wide, with runners four inches wide and standards six inches high; this sledge had no particle of metal in its construction, being composed entirely of wood, horn, and rawhide lashings. It weighed forty-eight pounds, and carried easily a load of a thousand pounds. After a two hundred and fifty mile trip round Inglefield Gulf, it made the long journey to the north and return to within two hundred miles of McCormick Bay, when it was abandoned for a lighter sledge. The second sledge was eleven feet by two, with three and one-half inch runners and six-inch standards. It weighed thirty-five pounds, and carried a load of over five hundred pounds. It broke down on the upward trip and was abandoned. The third sledge, made by Astrup, was ten feet by sixteen inches, with three-inch runners and two-inch standards; it weighed thirteen pounds, and carried a load of four hundred pounds. This sledge made the round trip of thirteen hundred miles, though carrying a load for only about eight hundred miles.

The result of this extended practical experience with sledges has been to show me that my previous ideas as to the great superiority of the toboggan type of sledge for inland-ice work (ideas gained during my reconnoissance in 1886, east of Disko Bay) were erroneous, and that the sledge with broad runners

and standards is *the* sledge.   Also, that the wear upon the
runners is practically *nil*, and that shoes of steel or ivory are
not only useless, but actually increase the tractive resistance.

Of even greater importance to our successful progress during the inland-ice journey than our sledges were the ski, or
Norwegian snow-skates.   Valuable as are the Indian snow-
shoes for Arctic work, the ski far surpass them in speed, ease
of locomotion, and reduced chances of chafing or straining the
feet.   On the upward journey I alternated between the snow-
shoes and the ski, but while descending the northern ice-slope
I had the misfortune to break one of the ski, and on the return trip was obliged to use the snow-shoes only.   Astrup
used ski entirely from start to finish.

I am satisfied that the only material for the clothing of men
traveling upon the inland ice is fur, and that the man who
dispenses with it adds to the weight he has to carry, and
compels himself to endure serious drafts upon his vitality, to
say nothing of deliberately choosing discomfort instead of comfort.   The great objection urged against fur clothing is that,
allowing the evaporation from the body no opportunity to escape, the clothing beneath it gets saturated while the wearer is
at work, and then, when he ceases, he becomes thoroughly
chilled.   This trouble is, in my opinion, due entirely to inex-
perience and ignorance of how to use the fur clothing.   It was
a part of my plan to obtain the material for my fur clothing
and sleeping-bags in the Whale Sound region, and I was entirely successful in so doing.   My boys shot the deer, the

skins were stretched and dried in Redcliffe, I devised and cut the patterns for the suits and sleeping-bags, and the native women sewed them.   As a result of my study of the Eskimo clothing and its use, I adopted it almost *literatim*, and my complete wardrobe consisted of a hooded deerskin coat weighing five and one-fourth pounds, a hooded sealskin coat weighing two and one-half pounds, a pair of dogskin knee-trousers weighing three pounds nine ounces, sealskin boots with woolen socks and fur soles, weighing two pounds, and an undershirt; total, about thirteen pounds.   With various combinations of this outfit, I could keep perfectly warm and yet not get into a perspiration, in temperatures from $+40°$ F. to $-50°$, whether at rest, or walking, or pulling upon a sledge.

The deerskin coat, with the trousers, footgear, and under-shirt, weighed eleven and one-fourth pounds, or about the same as an ordinary winter business suit, including shoes, underwear, etc., but not the overcoat.   In this costume, with the fur inside and the drawstrings at waist, wrists, knees, and face pulled tight, I have seated myself upon the great ice-cap four thousand feet above the sea with the thermometer at $-38°$, the wind blowing so that I could scarcely stand against it, and with back to the wind have eaten my lunch leisurely and in comfort; then, stretching myself at full length for a few moments, have listened to the fierce hiss of the snow driving past me with the same pleasurable sensation that, seated beside the glowing grate, we listen to the roar of the rain upon the roof.

Our sleeping-bags, also of the winter coat of the deer, with the fur inside, were, I think, the lightest and warmest ever used. In my own bag, weighing ten and one-fourth pounds, I have slept comfortably out upon the open snow, with no shelter whatever and the thermometer at $-41°$, wearing inside the bag only undergarments. During the inland-ice journey, throughout which the temperature was never more than a degree or two below zero, our sleeping-bags were discarded, our fur clothing being ample protection for us when asleep, even though I carried no tent.

While the variety of food was not as great as it has been on some other expeditions, I doubt if any party ever had more healthy or nutritious fare. A carefully studied feature of my project was the entire dependence upon the game of the Whale Sound region for my meat supply; and though I took an abundance of tea, coffee, sugar, milk, flour, corn-meal, and evaporated fruits and vegetables, my canned meats were only sufficient to carry us over the period of installation, with a small supply for short sledge journeys. In this respect, as in others, my plans were fortunate of fulfilment, and we were always well supplied with venison. With fresh meat and fresh bread every day we could smile defiance at scurvy.

OTHER
COOPER SQUARE PRESS
TITLES OF INTEREST

THE NORTH POLE
Robert E. Peary
Foreword by
Theodore Roosevelt
New introduction by
Robert M. Bryce
472 pp., 110 b/w
illustrations
0-8154-1138-3
$22.95

MY ATTAINMENT OF THE POLE
Dr. Frederick A. Cook
New introduction by
Robert M. Bryce
624 pp., 52 b/w
illustrations
0-8154-1137-5
$22.95

ARCTIC EXPERIENCES
Aboard the Doomed *Polaris*
Expedition and Six Months
Adrift on an Ice-Floe
Captain George E. Tyson
New introduction by
Edward E. Leslie
504 pp., 78 b/w
illustrations
0-8154-1189-8
$24.95 cloth

ANTARCTICA
Firsthand Accounts of
Exploration and Endurance
Edited by Charles Neider
468 pp.
0-8154-1023-9
$18.95

A NEGRO EXPLORER
AT THE NORTH POLE
Matthew A. Henson
Preface by
Booker T. Washington
Foreword by Robert E.
Peary
New introduction by
Robert M. Bryce
272 pp., 6 b/w photos
0-8154-1125-1
$15.95

THE *KARLUK'S* LAST VOYAGE
An Epic of Death and
Survival in the Arctic
Captain Robert A. Bartlett
New introduction by
Edward E. Leslie
378 pp., 23 b/w photos, 3
maps
0-8154-1124-3
$18.95